Preventing Legionellosis

Preventing Legionellosis

William F. McCoy

Publishing
LONDON • SEATTLE

Published by IWA Publishing, Alliance House, 12 Caxton Street, London SW1H 0QS, UK

Telephone: +44 (0) 20 7654 5500; Fax: +44 (0) 20 7654 5555; Email: publications@iwap.co.uk
Web: www.iwapublishing.com

First published 2005

Typeset by Gray Publishing, Tunbridge Wells, UK
Index prepared by Indexing Specialists (UK) Ltd

British Library Cataloguing in Publication Data
A CIP catalogue record for this book is available from the British Library

Library of Congress Cataloging-in-Publication Data
A catalog record for this book is available from the Library of Congress

ISBN 1 84339 094 9

Contents

Preface

Legionellosis has killed or severely debilitated at least a quarter million people since discovery of its cause and how to prevent it. Hundreds of thousands of people are newly afflicted each year. Successfully treating the acute disease is not good enough because survivors suffer long-term debilitating effects.

Most victims of legionellosis were exposed to improperly maintained building water systems. Most legionellosis cases are fundamentally the result of inadequate water treatment and poor water system management. Most cases are therefore preventable.

Legionellosis is the most serious form of pneumonia that people can get from exposure to water systems. Many thousands of cases could have been prevented by application of practical scientifically-proven hazard analysis and control principles.

Today, legionellosis still causes tremendous human suffering because:

- Public concern is not proportional to actual risk
- Codes of Practice and Standards are not adequate or not implemented
- Scientifically-based hazard analysis and control is not implemented
- Commercial conflicts of interest impede good practice.

Acknowledgements

The author thanks Jeffrey Cirillo, PhD (University of Nebraska), Rafael Garduno, PhD and Gary Faulkner, PhD (both from Dalhousie University) for their excellent micrographic contributions to Chapters 2 and 4.

Robert Brewster, RS, MPH, LEHP (Hazard Analysis and Critical Control Point (HACCP) Consultant and Educator, Innerspace Environmental Assessment, Inc., Elburn, IL) and Roy Costa, RS, MS (HACCP Consultant and Educator, Environ Health Associates, DeLand, FL) gave many helpful comments and suggestions about hazard analysis and control systems used throughout the book.

Thomas W. Armstrong, CIH (ExxonMobil Biomedical Sciences), is acknowledged for discussions about legionellosis incubation period (Section 3.1.2). Dr. Tom Makin, PhD (The Royal Liverpool University Hospital) is gratefully acknowledged for his interpretation of the radiographs in Figure 3.2.

Dr. Matthew Moore, MD, MPH, LCDR, USPHS (Medical Epidemiologist, National Center for Infectious Diseases, Centers for Disease Control and Prevention, Atlanta, GA) gave helpful comments and discussion about the epidemiology of the disease (Chapter 4). Larry Gess, Phigenics, LLC, is gratefully acknowledged for Figure 4.1.

For Chapter 7, Hazard Control, Tim Jones (Safepurge, Inc.), Steve Schira (LiquiTech, Inc.), Rodney Herrington and Katie Bolek (MIOX Corp.), Mike St. Germain and Noel Twyman (ProMinent Fluid Controls Corp.), James Furlong (Baltimore Aircoil Company), and Dave Geary (D.F. Geary Consultants, LLC) are acknowledged for helpful discussions, captions, and artwork.

The author especially thanks Barry Fields, PhD (Centers for Disease Control and Prevention, Atlanta, GA), Richard Miller, PhD (University of Louisville, Louisville, KY) and Janet Stout, PhD (VA Medical Center, Pittsburgh, PA) for many helpful comments and suggestions about preventing legionellosis.

Clive Broadbent, PE (Consulting Engineer, Canberra, Australia) and Richard Bentham, PhD (Flinders University of South Australia, Adelaide, Australia) are gratefully acknowledged for perspective about legionellosis Down Under.

Denis Green and the many people who actively communicate via internet discussion groups were sources of information regarding the tremendously high cost of legionellosis in the lives of thousands of victims and their families.

http://groups.yahoo.com/group/legionnaires_disease/
http://health.groups.yahoo.com/group/Legionnaires_diseaze_support_group/
http://www.q-net.net.au/~legion/
http://www.multiline.com.au/~mg
http://members.dodo.net.au/~jamgreen/legionellose_france.index.htm
http://home.iprimus.com.au/matgreen/Legionella_Spain.html

PART ONE: Legionellosis

1

Cost

The cost of legionellosis can be estimated in monetary terms but to truly understand, it is necessary to listen to victims describe their personal losses.

1.1 COST ESTIMATE

The conservatively calculated annual cost of legionellosis in America is at least $2 billion, as follows. Community-acquired pneumonia (CAP) is a common problem in emergency medicine (Stäuble *et al.* 2001). Incidence estimates are that about 2–6% of CAP cases requiring hospitalization are caused by legionellosis (Bartlett 1998). Overall incidence of CAP caused by legionellosis can be as high as 15% (Mulazimoglu 2001). Since about 485,000 cases of CAP require medical attention in the USA each year (Marston *et al.* 1997) and the direct healthcare cost of CAP in America is about $10 billion (Lave *et al.* 1999), therefore the calculated direct healthcare annual cost of community-acquired legionellosis is estimated at between $200 and $600 million. But the indirect cost of legionellosis is far greater because about $12 of expenses are incurred from lost time/lost productivity for every $1 of direct healthcare expense due to CAP (Birnbaum 2000).

This is an underestimate of the cost. Legionellosis is more likely to result in hospitalization with complications compared to other causes of CAP. Further underestimating the cost of legionellosis, this estimate does not include nosocomial (hospital-acquired) cases.

The direct healthcare cost of each nosocomial case of pneumonia in America was estimated to be $7000 (Bergmans 2004). This cost was extrapolated to estimate that the national annual direct healthcare cost of nosocomial legionellosis in America was at least $34 million (Squier 2005).

The direct healthcare cost of pneumonia treated in UK hospitals is about £1700–5100 per episode (Guest 1997). Several hundred legionellosis cases require medical attention in the UK each year (see Table 4.4).

Although the vast majority of legionellosis cases are never associated with outbreaks, those cases that are can be very expensive. Many US outbreaks have resulted in out-of-court settlements with payouts in the hundreds of thousands or millions of dollars. Usually as a condition of settlement, terms are not disclosed.

For instance, five family members and a friend contracted legionellosis from a hotel pool/spa in the US. The victims, including one who suffered in hospital for an entire month, sued for medical bills, lost wages, pain, suffering and punitive damages. During the discovery phase of the lawsuit, public records from every state in which the hotel chain operated revealed that 18 different health departments had shut down 25 unsafe pools/spas owned by the hotel chain. A substantial settlement was reached (terms not disclosed) before the case went to trial (personal communication, Lee Wallace, Attorney, Atlanta, GA).

A hospital-associated outbreak in 2005 killed one patient and resulted in costs of at least $328,000 for preventative measures, patient relocations and plumbing reconfigurations (*Iowa City Press-Citizen*, 6 May 2005).

The Melbourne Aquarium (Victoria, Australia) outbreak in 2000 (see 4.1.2.1) led to class action lawsuits brought on behalf of more than one hundred victims. The defendants were the owners, water treatment provider, builder, architect, contractors, equipment suppliers and insurers of the property. Sixteen people in the member group died before the lawsuit settled in 2004, four as a direct result of infection, the rest from complications (*The Australian*, Feb. 3, 2004, Nationwide News Pty. Ltd. and Nov. 26, 2004 Australian Associated Press Pty. Ltd.). The payout is believed to have been in the range of $AU40–70 million.

Of course, the cost of legionellosis is far greater than that which can be paid in dollars, pounds or euros.

1.2 PERSONAL ACCOUNTS OF ACUTE AND LONG-TERM ILLNESS

To really understand the cost of legionellosis, it is necessary to listen carefully to victims describe their experiences with the disease. The experiences described in these personal accounts can be multiplied by tens of thousands of times every year.

Roger from the UK writing in April 2005:

> *"I am a 64 year old male living in Bedford UK. I contracted this awful LD in December 2002 from a hotel shower in France whilst on a short visit, before which I was in perfect health and was leading a very active life. A week after I returned I started to feel ill with flu-like symptoms and after a few days became delirious dehydrated and had difficulty with breathing. I was then admitted to [name withheld] general hospital ICU and diagnosed with confirmed LD. My condition became rapidly worse and I lost consciousness and was put on a ventilator for two weeks. I then,*

after the treatment began to work, gradually improved and was discharged under home medical supervision after some six weeks in hospital. My time in hospital was a bit of a blur and all I remember is the awful pain and the terrible hallucinations. The flash backs are still to this day very distressing for me.

My recovery process has been long and slow and even after more than two years I still suffer from the following post LD effects of which I am now not hopeful of any improvement: (1) Difficulty with short term memory, (2) Lack of concentration, (3) Muscle and joint pain causing difficulty with walking, (4) Feeling tired all the time even though I am sleeping for 12 hours per day, (5) Very poor energy levels.

My life has been completely altered by this terrible disease and I am no longer able to function as I was before and I am continually in pain with serious mobility problems."

Hilary from the USA writing in April, 2005:

"In early September 2004, I developed a high fever, headache, and felt nauseated. This happened over the Labor Day weekend and the Tuesday following the Labor Day holiday I visited my family doctor where I was told that I had a gastrointestinal virus. At the time, I was a 47-year old healthy woman with no previous respiratory problems. After five days of running the fever (which sometimes was as high as 104F) my neighbor took me to the Emergency Room. Upon arriving at the ER, my blood pressure was 60/40. I had a series of diagnostic tests, and a chest x-ray revealed pneumonia. I had absolutely no respiratory symptoms prior to being admitted to the hospital. Upon admittance, my blood oxygen level was extremely low, but I did not have a cough or cold symptoms. In addition to the high fever, nausea and headache, I also had muscle pain, particularly in my legs. I was then admitted into the Intensive Care Unit where 12 hours later I developed sepsis and was intubated and placed in a drug-induced coma. I remained in the coma for 12 days and in ICU for 15 days. The last five days of hospitalization were spent on a medical floor.

When I was weaned from the coma, I was given gentle arm restraints so that I would not pull the feeding tube out or the breathing treatment therapy apparatus. I suffered horrifying hallucinations. I also became a bit crazy as well, accusing doctors and nurses, family members, and friends of raping me. I'm certain that the restraints had a lot to do with this. I thought I was in a hospital in Florida as at the time there were many hurricanes going on there and most likely, I heard this in the background on the television. I also had hallucinations of seeing my father and brother [both deceased] and wanted their telephone numbers once I was weaned off the medications from the coma.

The medication that most likely saved me was Levo [levofloxacin]. I believe that Levo was also the medication that made me lose my sense of reality. I credit the Infectious Disease doctor who jumped in right away to do the Legionnaires' disease testing. His name is [withheld]. I also credit the critical care/pulmonary doctors [names withheld] and [name withheld] a nurse practitioner, not to mention all the wonderful nursing staff.

There were about five days where my life was touch and go and I can't even imagine what was going on through my family member's and friend's minds. I know that a lot of tears were shed. I do remember that when I became more alert and family/friends visited me, they would cry with happiness knowing that I was turning the corner.

When I returned home, my husband arranged care for me by family/friends until he returned from work in the late afternoon. For the first two weeks, I also had a home nurse visit to check my vital signs, in addition to a physical therapist who gave me strengthening exercises to do. The home nurses also treated my wounds. I had very deep scars on my arms due to my veins collapsing from the IV lines. The scars are still there but fading. After five weeks at home, I was given permission to drive and started physical therapy on an outpatient basis. Four weeks upon discharge I had blood counts and chest X-ray and a follow-up appointment with the Pulmonary Doctors. I also had a Pulmonary Function Test which revealed normal results. I never needed any breathing treatments once I was discharged from the hospital.

It has now been seven months since I was so dreadfully sick. I still am recovering though and lack the energy and stamina I had prior to becoming ill. I also have had some problems with my vision (but this could be just an age-problem).

The absolute worst part of recovery has been remembering the hallucinations and fearing that this will happen again. I live with memory and fear daily.

I have no idea how I contracted Legionnaires' disease. I was at a beach home with my family two weeks prior to getting ill. No one else got sick. In my opinion, I think that those with severe pneumonia should be tested for the bacteria. It is my understanding that many hospitals do not have the testing materials and many cases are not diagnosed."

Phil in the USA writing in April 2005:

"I had Legionnaires' disease in October 2003 and was in a coma for 18 days. I was 47 years old at the time. While I was ill, my lungs filled with fluid, kidneys, liver failed, and my heart rate went from the floor to the roof, and a fever of 104.9F. I was on a ventilator, and had a feeding tube. I was in the hospital for about 25 days and lost 65 lbs. I was unable to walk when I left the hospital and spent two weeks in rehab to learn to walk again. I returned to work in January 2004 and have been working ever since. But I am having problems with the strength in my legs because they some times get weak on me and they seen to retain fluids at times. Has any one else after recovery had problems with their legs?"

Diane from the USA writing in April 2005:

"My husband contracted Legionnaires' disease in August 2003. He was 54 at the time and will be turning 56 this month. A day after we returned from a trip to the Bahamas, my husband had flu like symptoms. He was achy, had diarrhea, vomiting and a severe headache. Over the next few days the symptoms became worse, along with severe chills and fever. I took him to the emergency room of [name withheld]. He had a temperature of 103F, they took some blood tests and diagnosed him with a bacterial infection believing he may have drank contaminated water in the Bahamas. They sent him home with a prescription of Cipro, an antibiotic. The next day we saw no improvement, so I called the hospital ER and was told it would take a few days before he felt better. The day after that he became deathly ill. He started hallucinating and was literally seeing things that weren't there. I rushed him back to the ER. His oxygen level was very low. They did x-rays which showed severe pneumonia in both lungs and blood work indicated he was having or recently had a heart attack. I was told that night that he was in critical condition. They were testing him for SARS, West Nile Virus, etc. because we had been out of the country. The next day they told me he tested positive for Legionnaires' disease. He spent 11 days on a respirator. Fortunately, his kidneys didn't fail, as I've read that can happen with Legionnaires' disease. My husband spent 3 weeks in the hospital. After he was removed from the respirator he was transferred to [another hospital, name withheld]. He received excellent care there.

The days my husband spent on the respirator were horrible. The first few days he was heavily sedated but after that he was aware of everything. He was writing us notes asking if he was dying and we reassured him he would recover. We all researched the disease on the Internet to learn as much as we could. The doctors could tell us how it was contracted and treated but had very little information on the recovery. The stress took a toll on all of us. We have a large extended family and many friends who helped us through those first few weeks. But I have to say it was one of the most stressful times of our lives.

We believe my husband contracted Legionnaires' disease in a hot tub at the hotel where we stayed in the Bahamas. He had a compromised immune system, which is why he was susceptible to contracting the disease. He has rheumatoid arthritis.

It has been 20 months, yet he still suffers from lingering effects. He has some short term memory loss along with problems with math and numbers. He gets extremely tired. He was unable to return to work and is on permanent disability. We are thankful he survived this disease, but I don't think people are aware of just how devastating it can be."

Jacek from Poland writing in April 2005:

"I received a diagnosis of confirmed legionellosis at the Pneumology Hospital [name withheld] in Poland. I was in hospital for 5 weeks but the last two were mainly sitting at home and visiting hospital for tests.

I believe I contracted LD while travelling in Hong Kong and China, last June [not confirmed]. Weeks before the travel, I had bad influenza. I ignored it at the beginning – and then – I had serious problems to recover from it. Just 2–3 days before the travel to Asia, I finished antibiotics therapy.

Due to risk of tropical diseases, I took some vaccines (the doctor had refused first, as I had still not fully recovered from influenza, but I insisted, and finally he agreed. Later on, these facts have been considered the reason for the weakness of my immunology system.

I started to feel ill on the first day I returned to Poland. The symptoms were exactly like influenza: Headache, muscle pain, general discomfort. When I measured temperature (ca 39 deg C), then I realized that something was seriously wrong. The first idea was SARS, as some cases occurred at that time in China.

Next day morning, my wife took me to hospital – they let me stay for tests and this is how LD was diagnosed.

I strongly believe that my trip to China trip influenced the physicians to do very swift and accurate medical tests and arrive at a diagnosis and treatment."

Deidre from Ireland writing in April 2005:

"I was diagnosed [with confirmed legionellosis] on 30th May 2002 having been admitted to hospital with pneumonia-like symptoms late on the 28th. I was 34 years of age at that time.

I was admitted with a diagnosis of pneumonia and was being treated with oxygen, anti-biotic and steroids. During the night I deteriorated rapidly and was admitted to ICU on the following morning in which I remained for nearly eight weeks. During this time I went into multi-organ failure – lungs obviously with some hypoxia, liver, kidneys requiring dialysis, blood clotting failure resulting in anemia requiring blood transfusions. I had a naso-gastric tube and catheter fitted. At that stage my notes record that I was unlikely to survive but that nothing further could be added to my care. Also during this time I was nursed in a prone position to increase my lung capacity.

My sense of that time was one of being trapped in a book I knew the ending to but that I was being made to read again and again.

When I did come around I was unable to move and I do mean unable *– it was a case of blink once for yes and two for no. I also had a trachy tube so was unable to communicate in any real way with those around me – awful. The worst thing about the trachy was the seemingly never ending suction and when the cuff was inflated real difficulty in breathing as if through tar. The "T" piece on the trachy fell of on regularly which was terrifying because nobody told me that when this happened I wouldn't die. The complete lack of control was also an issue.*

Once I came 'round I found sleep to be very elusive. The constant reminders and advice to sleep was really frustrating. I felt very alone and afraid all the time. Due to risk of exposure to further infection I was only allowed to see family all of whom are scattered around the country.

The physiotherapy was also very frustrating and frightening but I was very lucky to have someone who encouraged me as opposed to heavily challenging me. When they first got me to sit up I had a sea-sickness feeling as all my organs fell back into place. Being sick with the trachy was no fun!

It took a while for me to master walking and talking – when I came home from hospital I wasn't able to stand on my own so there was no question about the need for rehabilitative care. I eventually made contact with a Brain Injury Team and through them accessed physiotherapy, occupational therapy, memory work, psychological support etc.

It took months of therapy to get all my movement back – every joint ached and I mean every joint!! My energy levels were very low and indeed still need to be watched as I now burn out very easily. I work three days a week and feel that it is all I am really capable of. My memory is atrocious and I am currently working with a neuropsychologist and neurologist. All of these supports have been located and accessed by myself.

I believe I contracted [legionellosis] from my own shower which had not been used for some time as I had been with my parents. I was also a member of a leisure club but no tests were ever done as I was an isolated case. [The actual source of this infection was not determined].

Another factor I find difficult is not being able to let go of what has happened – to me it should make a difference that it did! Maybe that's why things like this [the opportunity to tell the story so that others might not suffer as much] are cathartic for me."

1.3 PERSONAL INJURY, CRIMINAL NEGLIGENCE, AND CORPORATE MANSLAUGHTER

John Silivanch contracted legionellosis from a whirlpool spa while on the cruise ship *Horizon* (Celebrity Cruises, Inc.) in 1994. In September 2001, a US Magistrate Judge in New York federal court upheld a $9.6 million jury award for personal injury from legionellosis and fraud by rejecting all objections regarding negligence, punitive damages, and loss of society (*John and Joyce Silivanch v. Celebrity Cruises Inc., et al*, Master File No. 94 Civ. 5270, Southern District, NY).

The jury awarded the plaintiffs $7 million in punitive damages. The jury also awarded John Silivanch $110,000 for medical expenses, $1.35 million for lost earnings, and $900,000 for pain and suffering. Joyce Silivanch was awarded $300,000 for loss of society (Mealey 2001).

On the matter of negligence, the US Federal Court judge wrote:

> *"... there is evidence that the connection between operation of sand filters and the transmission of disease was sufficiently well-known to satisfy the foreseeability requirement."*

On the matter of punitive damages, he wrote:

> *"Punitive damages may be awarded where a defendant's conduct is intentional, wanton, and reckless or constitutes gross negligence."*

Regarding pain and suffering, he wrote:

> *"John Silivanch suffered brain damage that has rendered him unemployable, impedes his social relationships, disrupts his orientation as to time and place and inhibits his ability to learn. His condition is permanent and is unlikely to improve. In light of this evidence, the jury's award was not excessive."*

On the matter of loss of society, the Judge wrote on behalf of Mrs. Silivanch:

> *"... loss of society damages compensate for the loss of a broad range of mutual benefits each family member receives from the other's continued existence, including love, affection, care, attention, companionship, comfort and protection."*

In a UK court on 8 April 2005, Gillian Beckingham testified in her defense that she was not responsible for the deaths of seven people from legionellosis. As head of the department for building design services, she stood accused of corporate manslaughter. She was accused of being principally responsible for canceling the cooling water system maintenance contract at the Forum 28 Arts Centre in Barrow-in-Furness, Cumbria which consequently sprayed *Legionella*-contaminated water droplets into the atmosphere breathed in by passers-by. One hundred and seventy-two people were afflicted with confirmed legionellosis in July and August, 2002. The now deceased victims of the outbreak, all from Barrow, were June Miles, 56, Richard Macaulay, 88; Georgina Somerville, 54, Harriet Low, 74, Wendy Millburn, 56, Elizabeth Dixon, 80, and Christine Merewood, 55 (reported by Helen Carter, Saturday, March 12, 2005, The Guardian, UK). Ms. Beckingham was found guilty of breaching Health and Safety law but the jury was dismissed when it failed to reach verdicts after 23 hours of deliberation on the seven charges of manslaughter. A guilty verdict could have carried a sentence of life imprisonment. The families of the victims demanded a retrial. On April 24, 2005 Alistair Webster QC, prosecuting, said the Crown Prosecution Service (CPS) had decided that Ms. Beckingham should face a retrial. Mr. Webster told the court: "Careful consideration has been given at a senior level within the CPS and we are of the view that the public interest requires a retrial in this case."

Earlier in 2005, the entire Barrow Borough Council stood accused for the Barrow-in-Furness outbreak but was acquitted of manslaughter charges on a point of law in March 2005. This precedent-setting case was the first time corporate manslaughter charges have been brought against a local authority. However, the Council was found guilty of breaching Health and Safety Law.

Failure to control legionellosis in the UK may result in prosecution if a water system is found to be the source of health risks or there has been no assessment of that risk consistent with the 1999 Control of Substances Hazardous to Health (COSHH) Regulation and if necessary precautions to control the risks are not at least equivalent to the Approved Code of Practice and Guidance, Legionnaires' disease: the control of legionella bacteria in water systems (HSC 2000; also see Chapter 5).

In cases where there is no death as a result of an outbreak, prosecutions are brought against an individual or company with fines upon conviction in the Magistrates or Crown Court. Where death has occurred as a result of confirmed legionellosis, the authorities are under pressure to secure conviction for manslaughter. In such a case, the prosecution must establish that the defendant:

- owed a duty of care,
- breached his/her duty of care,
- the breach was in fact the cause of death,
- the breach can be considered gross negligence.

Gross negligence is:

- indifference or lack of foresight to an obvious risk of injury,
- a determination nevertheless to accept or avoid that risk,
- an intention to avoid it (the risk) with a high degree of negligence.

In the UK, if there has been no adequate risk assessment (see Chapters 5 and 6 for perspective on risk assessment) or no proper control measures taken following such a risk assessment, a defendant is open to accusation and prosecution of corporate manslaughter. Further, any person (or company of persons) whom the authorities (i.e., the police) consider guilty of gross negligence may be prosecuted. Recent prosecutions have now occurred against directors, managers, and even maintenance staff.

The current UK law in this area is complicated. Proposed reforms to the law would lead to even more aggressive prosecutions. The present government promised in 1997 to reform the law governing corporate responsibility and to create a new criminal offence called "corporate killing." A draft Corporate Manslaughter Bill was announced in November 2004 by the Queen. It is expected that the bill will go to Parliament early in the next term unless there is a change in Government at the next general election (Rose 2005). Unlimited fines, disqualification of directors, freezing company assets, and expensive compulsory remedial work are some of the penalties facing defendant individuals and organizations if the bill becomes law.

2

Cause

Legionellosis is the condition of being infected by *Legionella* bacteria. Transmission is by aspiration (inhalation) of contaminated water, aerosolized water droplets, or infective particles. No secondary (person-to-person) transmission has ever been documented.

2.1 THE PATHOGEN

The natural habitat of the pathogen is parasitic to aquatic protist hosts. *Legionella* bacteria incidentally infect and grow within mammalian phagocytic cells including human alveolar macrophage (white blood cells) and also alveolar epithelia (the cells that line the walls of the "air sacs" in the lung).

2.1.1 Discovery

2.1.1.1 Pathogen

During and immediately subsequent to the 1976 American Legion convention in Philadelphia, serum and tissues from patients with Legionnaires' disease were examined at the Centers for Disease Control and Prevention in Atlanta, Georgia. The etiological agent of legionellosis was discovered the next year (McDade *et al.* 1977). The first isolation of the legionellosis bacillus was from infected guinea pigs. The bacteria were successfully cultured in yolk sacs of embryonated eggs. Yolk-sac cultures of the bacillus were used as antigen to make fluorescent antibodies. Serum specimens from more than 100 patients with the clinical criteria of legionellosis showed diagnostic increases in antibody titers compared to controls. Use of the fluorescent antibody probe for examination of stored serum specimens from two previously unsolved outbreaks of severe respiratory disease revealed high titers to the legionellosis bacillus in most patients. These results proved that legionellosis was caused by a Gram-negative bacterium and that it was responsible for widespread infections.

2.1.1.2 Host

Legionella was the first bacterial pathogen shown to grow inside common free-living amoebae (*Acanthamoeba* and *Naeglaria*) from aquatic environments (Rowbotham 1980). Today, many potentially pathogenic intracellular bacterial parasites of amoeba are associated with human diseases (Table 2.1). This list shows that abatement of bacterial pathogen hosts (the protozoa) from engineered water systems is fundamental to preventing legionellosis and potentially, other waterborne disease.

2.1.2 Cell biology

Legionellae are γ-proteobacteria. They are Gram-negative, non-acid fast, heterotrophic, aerobic, transitionally motile, catalase-positive rods (Figure 2.1).

Typically about 1 μm long and half as wide, they can also grow into long filamentous rods (WHO 2002). Pre-formed amino acids for carbon and energy sources and an astonishing amount of iron are required for growth. They do not oxidize or ferment carbohydrates and they do not form spores (Fields *et al.* 2002). Figures 2.2 and 2.3 show ultrastructural detail of *Legionella pneumophila*.

Although they survive outside this intracellular environment (Stout 1985), apparently this is transient in the search to find new hosts to infect because growth outside the host appears to be ecologically insignificant (Fields *et al.* 1989; Fields 1993; Abu Kwaik *et al.* 1998; Newsome *et al.* 1998; Atlas 1999).

Legionellae are sensitive to many environmental and antimicrobial factors. They do not survive above 60°C nor can they grow below 20°C. They are unable to proliferate in aquatic environments without their hosts or perhaps complex diverse biofilms (Surman *et al.* 2002). Outside their hosts and apart from protection in microbial biofilms (Costerton *et al.* 1995), they succumb to typical antimicrobials at typical use concentrations (Wireman *et al.* 1993; Walker *et al.* 2003). But they do seem somewhat more tolerant of certain non-oxidizing antimicrobials and completely tolerant, at typical use-doses, to certain other antimicrobials compared to typical Gram-negative aquatic bacteria (Kim *et al.* 2002).

2.1.2.1 Genetics

The entire genomic sequence of *L. pneumophila* has been determined for the strain derived from the original 1976 isolate. It consists of a single circular chromosome with 3,397,754 base pairs (Chien *et al.* 2004). The fully sequenced genome most closely related to *L. pneumophila* is the obligate intracellular pathogen *Coxiella*

Table 2.1 Bacterial parasites of *Acanthamoeba* and other aquatic protozoa.

Bacterial parasites	Human diseases	References
Afipia spp.	Unknown. Possible link to cat scratch disease, septic arthritis, bone marrow infections.	La Scola *et al.* 2000
Several novel spp. in the *Rickettsiales*	Endosymbionts of amoeba related to *Rickettsia* that cause infections transmitted by ticks and other insects.	Fritsche *et al.* 1999
Burkholderia psuedomallei and other spp.	Melioidosis (infections that can lead to pulmonary, bloodstream, and/or organ disease).	Inglis *et al.* 2000
Candidatus Odyssella thessalonicensis	Unknown. A newly described genus and species of obligate intracellular *Acanthamoeba* parasite.	Birtles *et al.* 2000
Chlamydia pneumoniae	Community-acquired pneumonia and/or bronchitis; possible association with atherosclerotic vascular disease and with asthma. Causative agent in Psittacosis.	Essig *et al.* 1997
Coxiella burnetti	Q-fever. Community-acquired pneumonia.	Maurin and Raoult 1999
Ehrlichia-like organisms	Unknown. Possible link to ehrlichosis (bacterial infections transmitted by ticks)	Michel *et al.* 1995
Francisella tularensis	Tularemia produces symptoms that can be confused with typhoid fever or other *Salmonella* infections, brucellosis, malaria, tuberculosis, Q-fever, psittacosis, or legionellosis.	Abd *et al.* 2003
Helicobacter pylori	Peptic ulcer disease (90% of all ulcers).	Winieck *et al.* 2002
Legionella spp.	Legionellosis (Legionnaires' disease).	Rowbotham 1980; Barker *et al.* 1992
Legionella-like amebic parasites (LLAP)	Unknown. Potentially causes community-acquired pneumonia.	Adeleke *et al.* 1996
Listeria monocytogenes	Listeriosis (food poisoning).	Ly and Müller 1990
Mycobacterium spp.	Bacteremia, multi-organ bacterial disease, pulmonary infections, especially in HIV/AIDS patients.	Steinert 2001
Neochlamydia harmannellae	Unknown. Potential *Chlamydia*-like intracellular pathogen.	Horn *et al.* 2000
Parachlamydia acanthamoeba	Possible community-acquired pneumonia	Greub and Raoult 2002
Simkania negevensis (in the Bacterial Order *Chlamydiales*)	Associated with bronchiolitis in infants and pneumonia in adults. *Chlamydia*-like intracellular pathogen.	Kahane *et al.* 2001

HIV: human immunodeficiency virus; AIDS: acquired immune deficiency syndrome.

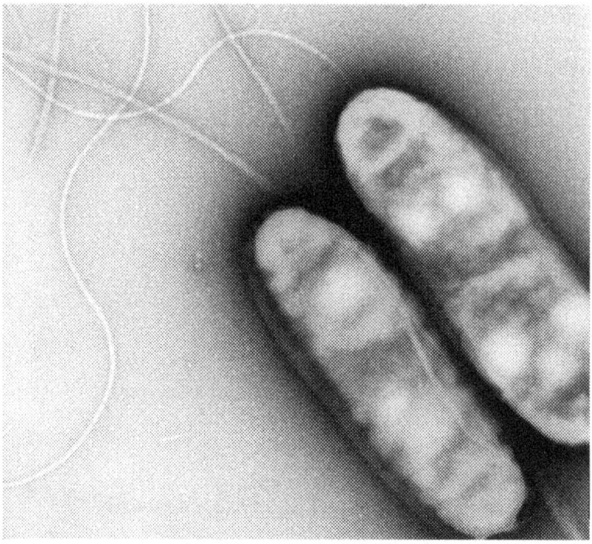

Figure 2.1 Negative stain of *Legionella pneumophila* serogroup 1. Bacteria were grown on buffered charcoal yeast extract (BCYE) agar and colonies picked, suspended in phosphate buffered saline (PBS) then overlaid onto carbon coated copper grids. The bacteria were then stained with uranyl acetate and visualized by transmission electron microscopy. A single polar flagellum can be seen protruding from the bacterium on the right. For size reference, the width of the flagellum is about 15 nm. Flagella usually have a wavy appearance; whereas, bundled conjugative pili appear relatively straight and rigid (image provided by Dr. Jeffrey D. Cirillo, University of Nebraska, USA).

burnetii, more that 40% of their genes are identical. About 60% of the genes in *Legionella* have homologs amongst other intracellular bacterial pathogens, such as *Coxiella*, *Salmonella*, *Chlamydia*, *Rickettsia*, *Brucella*, and *Mycobacterium*. So far, only *Legionella* and *Coxiella* are known to posses the genes referred to as the *dot/icm* (*d*efective for *o*rganelle *t*rafficking/*i*ntra*c*ellular *m*ultiplication) gene complex. Every *Legionella* mutant of the *dot/icm* family of genes is defective for evasion of the endocytic pathway and therefore, these mutants are ineffective in early survival after phagocytosis and do not grow inside the host (Swanson and Hammer 2000).

Legionellae possess genetic elements that impart an ability to infect and grow within phagocytic protist host cells, like amoeba, and also within mammalian phagocytes in the immune system, like human macrophage (Segal and Shuman 1999). The first of these genes discovered (Cianciatto *et al.* 1989) was the *m*acrophage *i*nfectivity *p*otentiator (*mip*). The *mip* gene and the Mip protein that it encodes are virulence-associated factors in *L. pneumophila*. Infections of guinea pig, mammalian phagocytic cells, and protozoan cells depend on a functional *mip* gene and Mip protein. This protein is profusely shed from the surface of *Legionella* inside phagosomes of the protist *Acanthamoeba* (Helbig *et al.* 2002). Although Mip is an active peptidyl-prolyl-*cis/trans* isomerase, this enzymatic function does not seem to be necessary in pathogenesis; the mode of action for the Mip virulence factor is unknown.

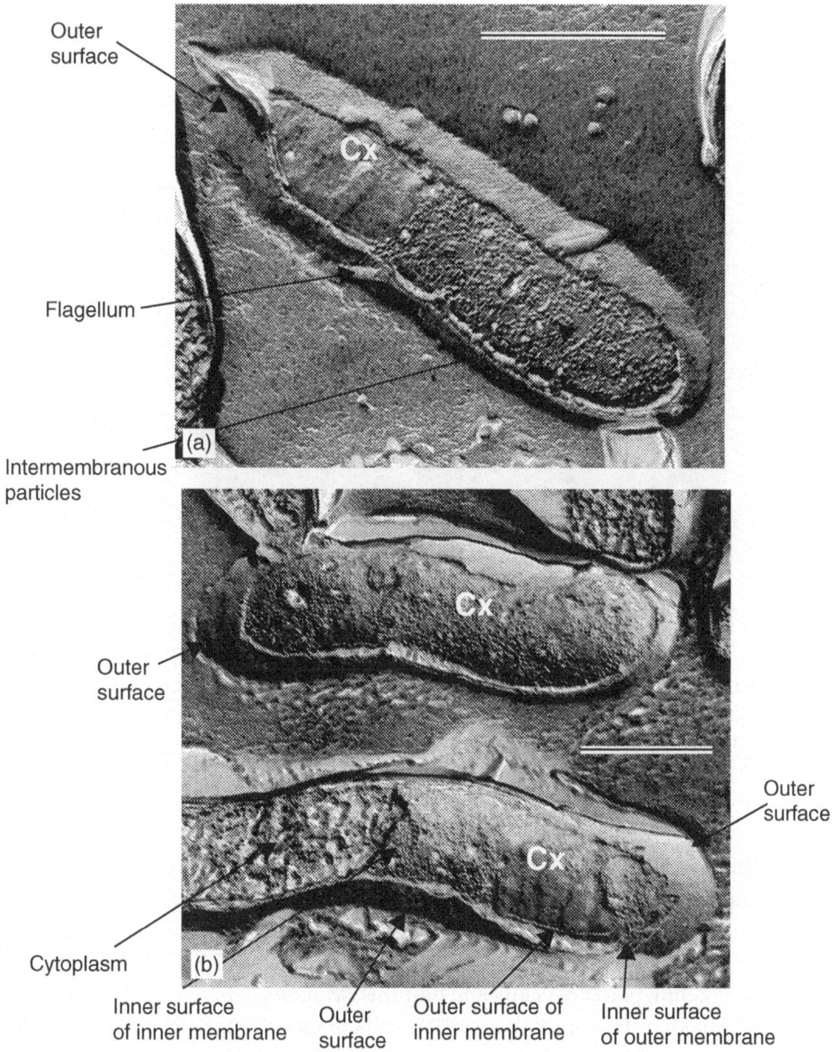

Figure 2.2 Freeze-fracture/etching micrographs of plate grown *L. pneumophila*. Several layers of the cell wall can be distinguished (as labeled). Micrograph (a) shows parts of a flagellum wrapped around the bacterial body. Shadows (darker areas) in the pictures should be interpreted as if the light source was at the top. Cx: convex. Bars = 0.5 μm. Micrographs by Gary Faulkner and Rafael Garduno, Dalhousie University, Halifax, Nova Scotia, Canada.

Legionellae in search of new hosts to infect present hair-like structures on their surfaces called pili that base in the inner membrane, extend through the periplasmic space and onward through the outer membrane into the extracellular environment (Stone and Kwaik 1998). Pili are used for attachment to surfaces and/ or transmission of genetic elements such as plasmids. *Legionella* process pilin

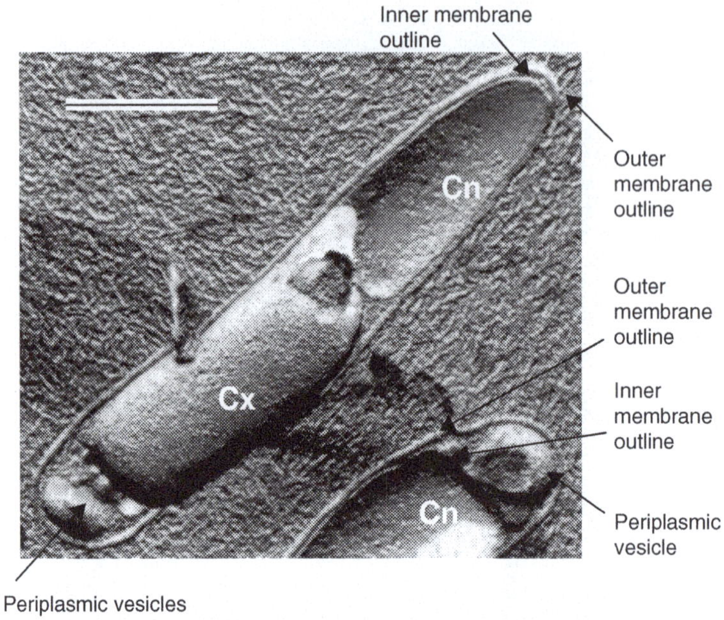

Figure 2.3 Freeze-fracture micrograph of *L. pneumophila* showing periplasmic vesicles, which are usually prominent at one cell pole. Shadows (darker areas) in the pictures should be interpreted as if the light source was at the top. Cx: convex; Cn: concave. Bar = 0.5 μm. Micrograph by Gary Faulkner and Rafael Garduno, Dalhousie University, Halifax, Nova Scotia, Canada.

protein subunits into adhesive type IV pili with prepilin peptidases, as do other Gram-negative pathogenic bacteria, such as *Pseudomonas aeruginosa* and *Neiserria gonorrhoeae*. Type IV pili are characterized by an adhesive hydrophobic tip and they are retractable. The *L. pneumophila* prepilin peptidase, protein PilD, is necessary for attachment to host cells and biofilms. Mutants deficient in the *pilD* gene are not as infective. The gene also appears to be necessary for efficient intracellular growth in macrophage and protozoa (Helbig *et al.* 2002).

2.1.2.2 Protein secretion

Legionella exploit the type IV protein secretion system for virulence (Nagal and Roy 2003). Type IV systems transport proteins or nucleic acid from the *dot/icm* gene complex across the bacterial cell wall directly into the host (Lee and Schneewind 2001). There are 24 known genes located in two separate loci on the bacterial chromosome that encode proteins secreted through the type IV machinery.

The inner membrane-bound PilD protein is also related to the type II protein secretion system common in many Gram-negative bacteria. The proteins these genes encode are necessary for intracellular growth of *Legionella* in their natural protist hosts and also in alveolar macrophage. They include lipases/esterases, phospholipase A, acid phosphatases, and proteases (Rossier and Cianciotto 2001;

Aragon *et al.* 2002; Rossier *et al.* 2002). Some of these proteins may be responsible for pathogenicity traits in Legionnaires' disease (see Section 7.2.2).

Attachment of *Legionella* to macrophage can occur at the major outer membrane protein (MOMP) which is encoded by *ompS.* The MOMP binds complement component C3 and thereby mediates uptake via the CR1 and CR3 macrophage receptors (Heuner *et al.* 2002). Another *Legionella* surface protein implicated in attachment is the heat shock protein (Hsp60) encoded by *htpB*. Hsp60 effects attachment and entry to HeLa epithelial cells (Faulkner and Garduño 2002). Since MOMP binding is complement dependent and is therefore mediated by the immune response, there appears to be no ecological role for this attachment mechanism in protist hosts. Likewise, Hsp60 binding also appears to be due to its antigenic attributes and an ecological role in protist infection is unknown.

2.1.2.3 Iron metabolism

Legionella require a lot of iron. To get them to grow in culture media, astonishing concentrations are necessary ($>10\ \mu M$) as ferric pyrophosphate, ferric citrate, or ferric chloride. Stationary phase liquid cultures present a dark brown color due to a pigment called homogentisic acid (HGA)-melanin that is iron dependent, hemolytic and may act as an extracellular sink for excess iron. Mutational analysis of iron assimilation deficiency proves that iron acquisition is necessary for intracellular parasitic growth of *Legionella*. To assimilate iron, they may produce as many as three siderophores (iron-carrying molecules) with biochemical evidence for a unique legiobactin and genetic evidence for hydroxamate-like and pyoverdin-like iron chelators. *Legionellae* can bind and use hemin and therefore, they may use heme-containing proteins as iron sources. *Legionella* may have the capacity to transport ferrous iron across its cell membrane suggesting that life is possible for them perhaps in a reduced environment. Fully oxidized ferric iron is certainly transported across and assimilated intracellularly. This impressive diversity in iron assimilation may account for *Legionella* survival in certain harsh aquatic habitats (Cianciatto *et al.* 2002).

The cell biology of *Legionella* is summarized schematically in Figure 2.4.

2.1.2.4 Taxonomy

There are at least 48 species of *Legionella* but five have been designated *Legionella*-like amoebic pathogens. All of them are potentially pathogenic. Although *L. pneumophila* serogroup 1 is thought to be the most virulent, it is first among many other potentially virulent serogroups. There are 15 serogroups of *L. pneumophila* and as many as 70 total among all the species in the genus. Two other genera have been proposed: *Fluoribacter* for blue–white fluorescing species, such as *L. bozemanii* and *Tatlockia* for the species *L. micadadei*.

Legionellae are phylogenetically close relatives of *C. burnetti*, the bacterial pathogen that causes Q-fever. These parasites have been shown to possess many interesting similarities in modes of survival within phagocytic cells (Maurin and Raoult 1999; Swanson and Fernandez-Moreia 2002).

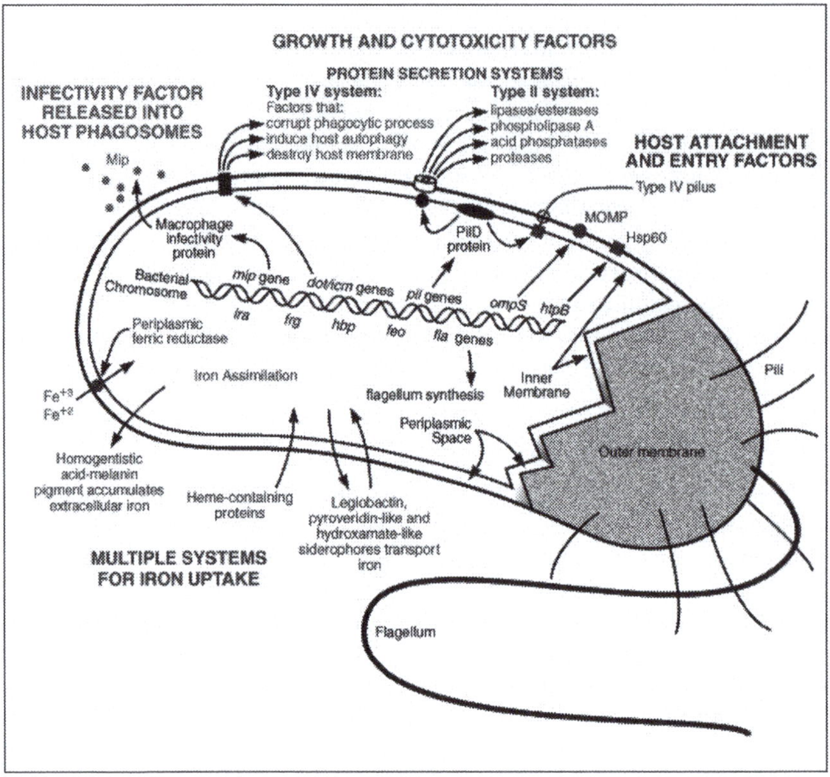

Figure 2.1 Summary of *Legionella* cell biology. The architecture of the cell wall is consistent with that of other Gram-negative bacteria. Virulence factors are encoded by the *dot/icm* genes, the *mip* gene, the *pil* genes, iron-uptake genes (*ira*, *frg*, *hbp*, *feo*), and for flagellum synthesis, the *fla* genes. Attachment to certain phagocytic hosts and to surfaces is effected by the type IV pilus, the MOMP, and the Hsp60. An important infectivity factor is the Mip protein. There are two protein secretion systems necessary for survival inside the host. Proteins secreted by the type IV system corrupt the phagocytic process and provide the pathogen an environment inside the host in which to grow and then later, to escape by forming pores in the host cell membrane. Proteins secreted by the type II system are necessary for growth within the corrupted phagosome. A uniquely diverse iron assimilation capability is necessary for intracellular growth. (As shown in *Microbial Waterborne Pathogens*, E. Cloete *et al.* (eds) ISBN: 1843390558; IWA Publishing 2004).

2.2 THE HOST

2.2.1 Protists

The amoebic and ciliated phyla (groups) of protozoa appear to be the most susceptible natural hosts to *Legionella* but very little is known about the range of this parasitic association; there may be many other undiscovered protist hosts.

Protists exhibit an enormous diversity in form and are mostly microscopic ranging in size from 10 to 200 μm. At least 60,000 species have been described but almost surely this is but a fraction of the total number of protist life forms. There are plant-like protists (the algae), fungus-like protists (the slime molds), and animal-like protists (the protozoa). There are four phyla of animal-like protists: amoebic (*Sarcodina*), ciliated (*Ciliophora*), non-ciliated (*Sporozoa*), and flagellates (*Mastigophora*). At least 14 genera of protozoa are certainly susceptible to infection by *Legionella*. The best-known protozoan hosts are *Acanthamoeba*, *Hartmannella*, *Naegleria*, and *Tetrahymena*. It is also clear that *Legionella* infect at least one fungus-like protist of the slime mold variety, *Dictyostelium discoideum* (Steinert *et al.* 2002). Infection occurs in the haploid amoebic form of this microorganism, which grows as separate, independent cells but then they interact to form multicellular structures when challenged by adverse conditions such as starvation. Individual cells signal each other by releasing the chemoattractant cyclic adenosine monophosphate (cAMP) and aggregate together by chemotaxis to form a mound that is surrounded by an extracellular matrix. Formation of a multicellular organism in this way is unique in nature.

Many protozoa make their livings by consuming bacteria, especially in biofilms (Barker 1992; Fields 1993; Watnick and Kolter 2000; Donlan 2002). They can be microscopically observed doing so in essentially all natural and engineered water systems. These grazers can quickly decimate bacterial biofilms and when they run out of biofilm, they form cysts to wait for biofilm redevelopment (Murga *et al.* 2001). *Legionellae* present a special problem because instead of being consumed along with other bacteria in the biofilm, these parasites can grow within and then overwhelm their protozoan hosts.

2.2.1.1 Infections of protists

Legionellae infect protists by corrupting the phagocytic process. In the early 1980s it was observed that they can enter the host by coiling phagocytosis and this has been confirmed (Bozue and Johnson 1996). *Legionella* end up in phagosomes after entering the host but the parasite inhibits early fusion with the lysosome mostly because of factors derived from the *dot/icm* gene complex. Bacterial lysis is thus avoided. Instead, *Legionella* causes the phagosome to associate with endoplasmic reticulum, exploit autophagy, and combine with lysosomes from the endocytic pathway. The pathogen, now in its ultrastructurally distinct replicative form, then prolifically multiplies within these acidified vesicles (Figure 2.5).

After only a few hours, thousands of lively motile bacteria can be observed to have arisen as a result of intracellular growth within the host protozoan. They swarm around voraciously until within as little as 48 h, the host cell becomes completely overwhelmed by the infection (Figure 2.6).

During these catastrophic biological events that ultimately lead to cell death, protozoa appear to attempt a clearing of the infection by expelling pathogen-laden vesicles from the cytoplasm (Figure 2.7). These small vesicles (about 5 μm diameter), are surrounded by membrane and are thought to be potentially infective in mammals since they can contain hundreds of virulent *Legionella* cells. They certainly appear to be relevant because transmission of the pathogen by this vehicle in the environment can occur (Berk *et al.* 1998).

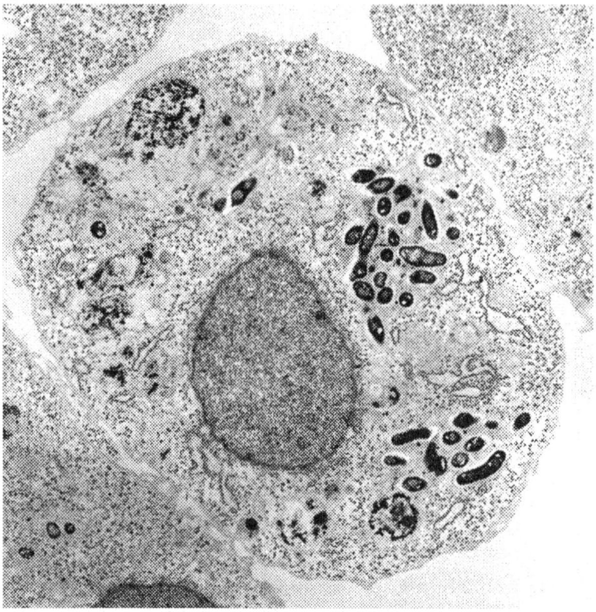

Figure 2.5 Transmission electron micrograph of environmental amoeba, *Acanthamoeba castellanii* strain Neff, infected with *L. pneumophila* serogroup 1. Multiple tight vacuoles can be seen in the perinuclear region of the amoeba. Micrograph by Dr. Jeffrey D. Cirillo, University of Nebraska, USA.

Other protists that are subject to infection by *Legionella* include members of the ciliate protozoa such as *Tetrahymena*. Having been consumed by the protist host, *Legionella* were observed in food vacuoles (Figure 2.8).

2.2.2 Mammalian phagocytes

The mammalian immune system includes specialized cells that protect against infectious disease by phagocytosis of pathogenic microorganisms. There are of course distinct differences between the phagocytic process of these cells and protozoan phagocytic cells such as amoeba. But the process is sufficiently similar to an extent that allows for infection, multiplication, and lysis by *Legionella*. Figure 2.9 shows the phagocytosis of *Legionella* by human phagocytic cells in culture.

Uptake of *Legionella* by white blood cells can be by coiling phagocytosis (Figure 2.10).

After uptake of *Legionella* in to the phagocyte, intracellular replication of the bacteria can be observed in vesicles closely associated with endoplasmic reticulum (Figure 2.11).

2.3 LIFE CYCLE

The parasitic life cycle of the *Legionellae* is bimodal with definable stages: host infectivity, intracellular growth, motility, destruction of the host, and release of

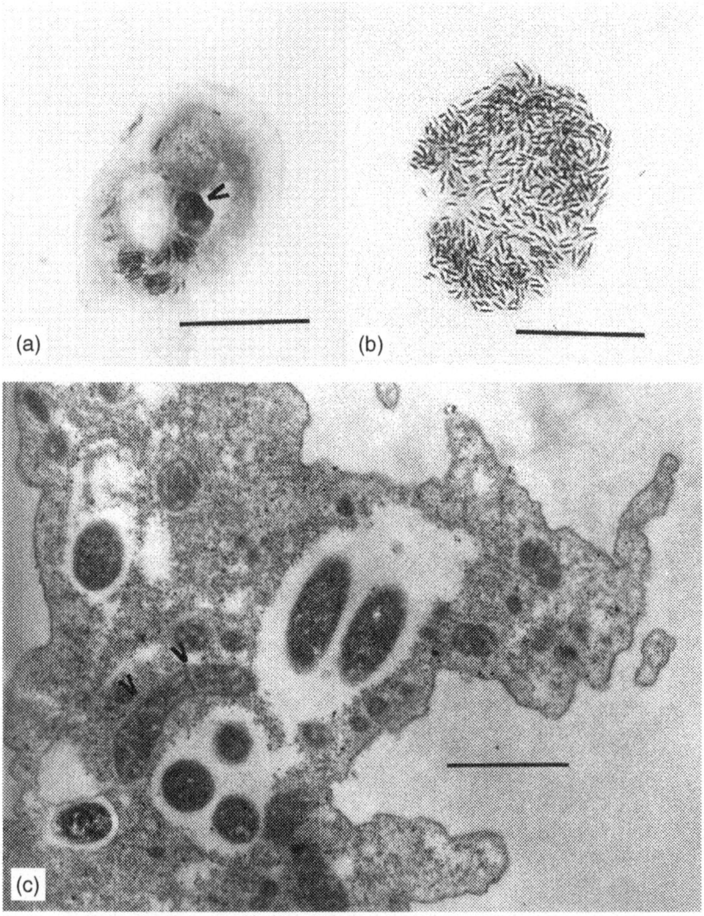

Figure 2.6 Giemsa stain showing the infection of an amoeba cell after 24 h (a) and 72 h (b) of incubation. Characteristics features of the amoeba such as the cell nucleus (arrowhead) were intact at 24 h but by 72 h, the host cell had lost intracellular structure and was beginning to lyse. Bar = 20 μm. By transmission electron microscopy (c), alignment of mitochondria to a vacuole-containing *Legionella* during early stages of the lytic cycle. Bar = 1 μm (Newsome *et al.* 1998. *Appl. Environ. Microbiol.* **64**(5): 1688). Reprinted with permission of the American Society of Microbiology.

progeny into the environment. Figure 2.12 provides a graphical summary of the *Legionella* life cycle in an aquatic environment.

Ultrastructurally distinct cell differentiation occurs during the *Legionella* life cycle (Figure 2.13). After intracellular replication, the pathogen differentiates into the infectious *m*ature *i*ntracellular *f*orm (MIF) that is motile and can survive prolonged starvation and harsh conditions when released back into the environment. Upon gaining entrance by infection to the intracellular parasitic environment of a

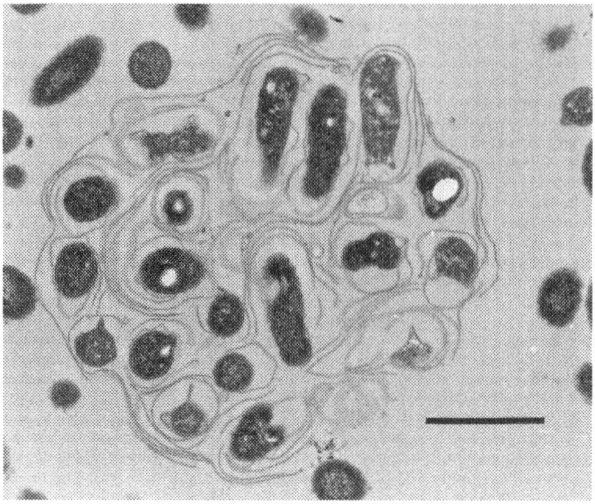

Figure 2.7 Transmission electron micrograph of a *Legionella*-laden vesicle expelled from the host cell *A. castellanii*. Bar = 1 μm (Berk *et al.* 1998. *Appl. Environ. Microbiol.* **64**(1): 279). Reprinted with permission of the American Society of Microbiology.

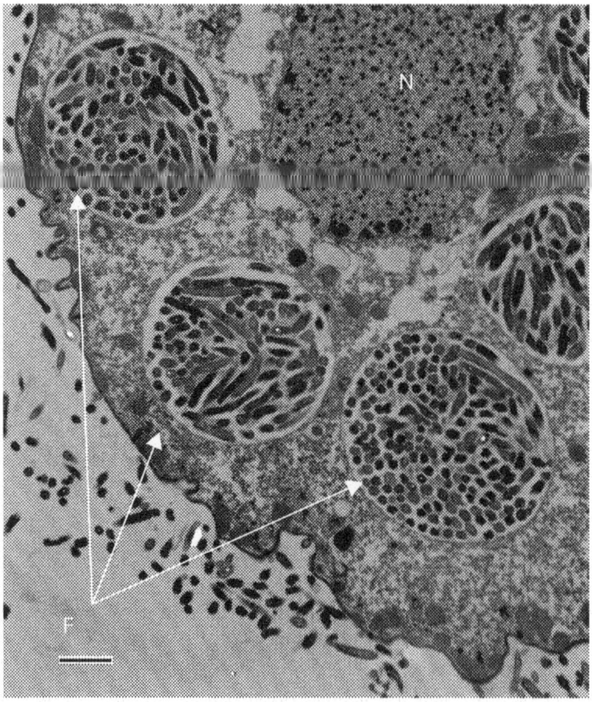

Figure 2.8 Thin section of the ciliate protozoan *Tetrahymena* (N: nucleus of the protist) that has ingested large numbers of *L. pneumophila*. Notice the independent vacuoles (F: food vacuoles) which are homogeneous in size and shape. Bar = 2 μm. Micrograph by Gary Faulkner and Rafael Garduno, Dalhousie University, Halifax, Nova Scotia, Canada.

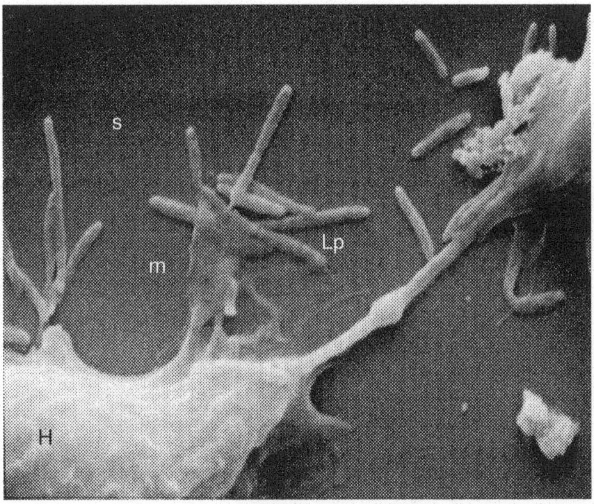

Figure 2.9 Scanning electron micrograph of human cell culture (HeLa) in active phagocytosis of *L. pneumophila* (Lp). Lp, Philadelphia-1 strain SVir was added to HeLa cells grown on glass coverslips and fixed 3 h after inoculation. s: surface of the glass coverslip; h: HeLa cell; m: membrane extension involved in the phagocytosis of Lp. Bar = 2 μm. Micrograph by Gary Faulkner and Rafael Garduno, Dalhousie University, Halifax, Nova Scotia, Canada.

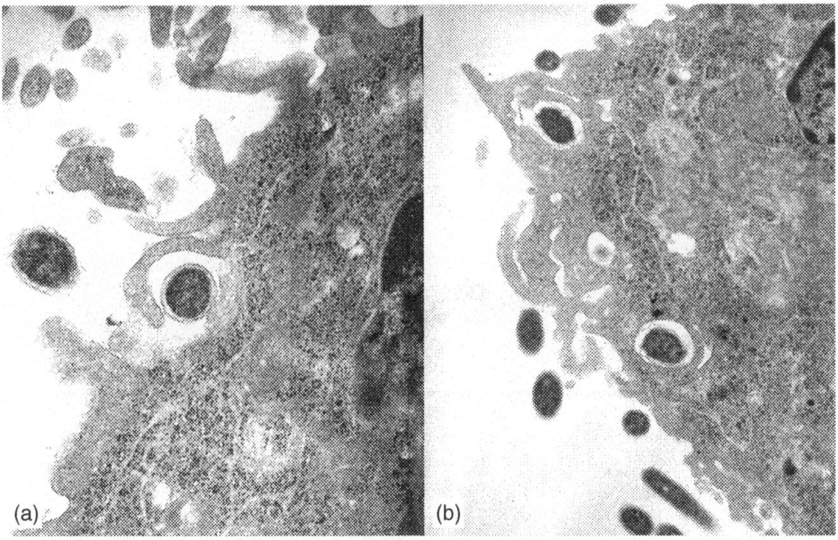

Figure 2.10 Transmission electron micrographs of coiling phagocytosis of *L. pneumophila* serogroup 1 into human macrophages (white blood cells). (a) Early during phagocytosis, bacteria adhere to the surface of the cell and an asymmetric philipodia is extended from the cell to wrap around the bacteria. (b) Upon uptake multiple membranes surround the bacteria in vacuoles. The membranes appear to fuse along the sides at multiple locations, resulting in a multiple membrane-lined compartments. Micrographs by Dr. Jeffrey D. Cirillo, University of Nebraska, USA.

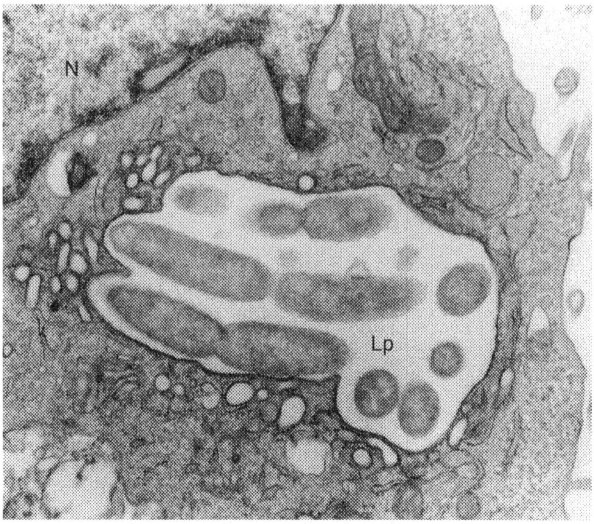

Figure 2.11 Transmission electron micrograph of an infected human cell (N: nucleus of host cell) with replicative *L. pneumophila* (Lp) inside a cell vacuole surrounded by numerous vesicles and portions of endoplasmic reticulum. Bar = 1 μm. Micrograph by Gary Faulkner and Rafael Garduno, Dalhousie University, Halifax, Nova Scotia, Canada.

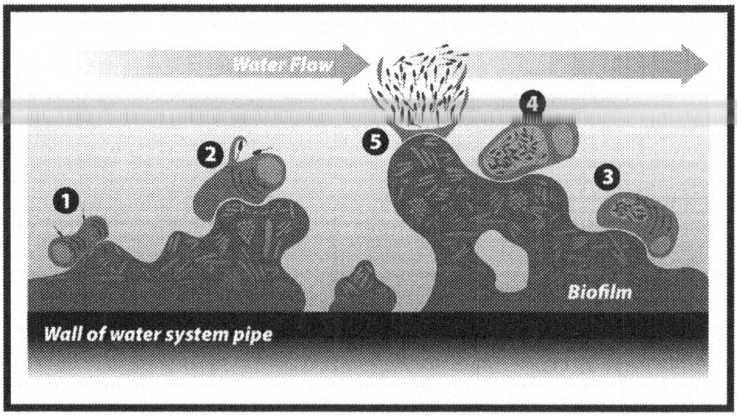

Figure 2.12 Life cycle of *Legionella* in an aquatic system. (1) Amoebae graze on biofilm consuming bacteria for food but are (2) infected by *Legionella* which, (3) grow intracellular in association with endoplasmic reticulum and then (4) overwhelm its protozoan host cell, differentiate into motile cells and (5) cause lysis of the host cell releasing hundreds or thousands of infective progeny in search of new hosts to infect.

new host, several ultrastructurally distinct intermediate forms differentiate finally into the delicate non-motile yet prolific replicative form of *Legionella*. Thus, it is clear that the *Legionella* life cycle is bimodal and highly pleomorphic (Faulkner and Garduño 2002).

Figure 2.13 These freeze-fracture micrographs show the striking difference between MIFs (a) and replicative forms (b) with respect to the slender shape of the replicative form as opposed to the often "chubby" shape of MIFs, the distribution of intramembranous particles (imp) which appear as bumps on the surface of the inner membrane and the periplasmic vesicles (p) in the replicative form. Micrographs by Gary Faulkner and Rafael Garduno, Dalhousie University, Halifax, Nova Scotia, Canada.

2.3.1 Infection

After entrance into the host phagosome, several *dot/icm* dependent factors immediately isolate the phagosome from its normal course; this allows *Legionella* to survive the first few minutes of infection because fusion of phagosome with lysosome is delayed. During the next several hours, the phagosome-enclosed parasite remains in lag phase, isolated from the lytic endosomal trafficking process and the pathogen becomes acid resistant (this acid resistance attribute is the basis for the selectivity step with HCl in the ISO analytical method, see Chapter 8, Validation and Verification). Factors encoded by *dot/icm* and secreted through type IV machinery in *L. pneumophila* cause the phagosome to associate with endoplasmic reticulum (Fields *et al.* 2002; Swanson and Bachman 2002).

2.3.2 Rapid intracellular growth

The next phase in the *Legionella* life cycle is characterized by rapid growth of the pathogen in corrupted phagosomes (Figures 2.14–2.17). It involves exploitation of autophagy in the host. Autophagy is a response to nutrient deprivation and it is ubiquitous among eukaryotic cells. A distinguishing feature of autophagy is the dynamic rearrangement of cellular membrane to sequester cytosol and organelles for delivery to the lysosome or vacuole wherein the contents are degraded

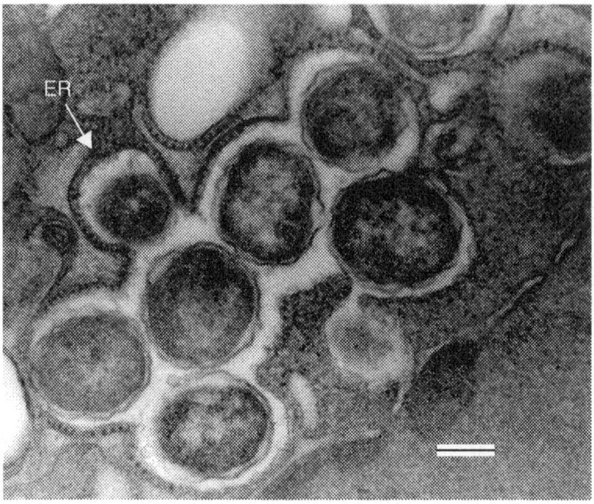

Figure 2.14 Transmission electron micrograph of an infected HeLa cell-containing replicative *L. pneumophila* showing the typical Gram-negative cell walls. In this case the bacteria are inside a vacuole that is decorated with rough endoplasmic reticulum (ER). Notice how well the vacuolar membrane follows the contour of the contained bacteria, suggesting a close molecular interaction between the bacterial cell surface and the vacuolar membrane. Bar = 0.2 μm. Micrograph by Gary Faulkner and Rafael Garduno, Dalhousie University, Halifax, Nova Scotia, Canada.

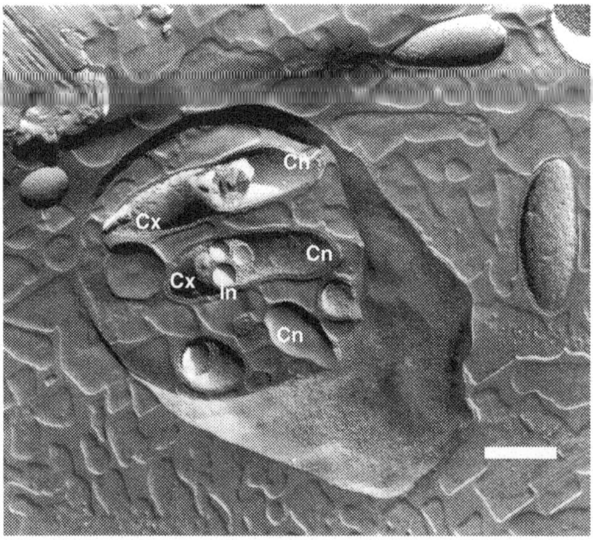

Figure 2.15 Freeze-fracture/etching micrograph of infected HeLa cells showing a few MIFs inside a vacuole. Notice the spherical cytoplasmic inclusions, In, which have a concave shape, Cn. Shadows are to be interpreted as if a light source was situated in the top right side of the picture. Therefore, the fracture plane where the MIFs emerge is located in a depression, Cn rather than in a protrusion, Cx. Bar = 0.5 μm. Micrograph by Gary Faulkner and Rafael Garduno, Dalhousie University, Halifax, Nova Scotia, Canada.

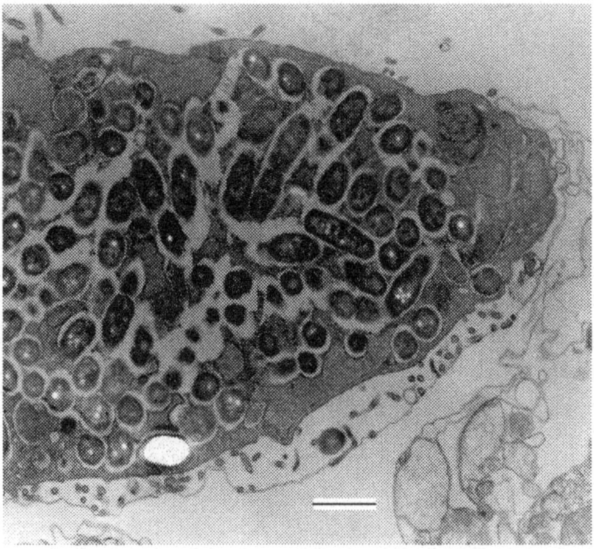

Figure 2.16 HeLa cell packed full of *L. pneumophila* (Lp) reaching the end of the replication cycle. Notice that the cytoplasm of the HeLa cell is still electron dense and that the membrane of the replicative vacuole still follows the contour of the contained bacteria. At this phase of the growth cycle the replicative vacuole is structurally complex, and it is possible to identify many bacterial morphological intermediates as Lp differentiates from a replicative form into a MIF. Bar = 1 μm. Micrograph by Gary Faulkner and Rafael Garduno, Dalhousie University, Halifax, Nova Scotia, Canada.

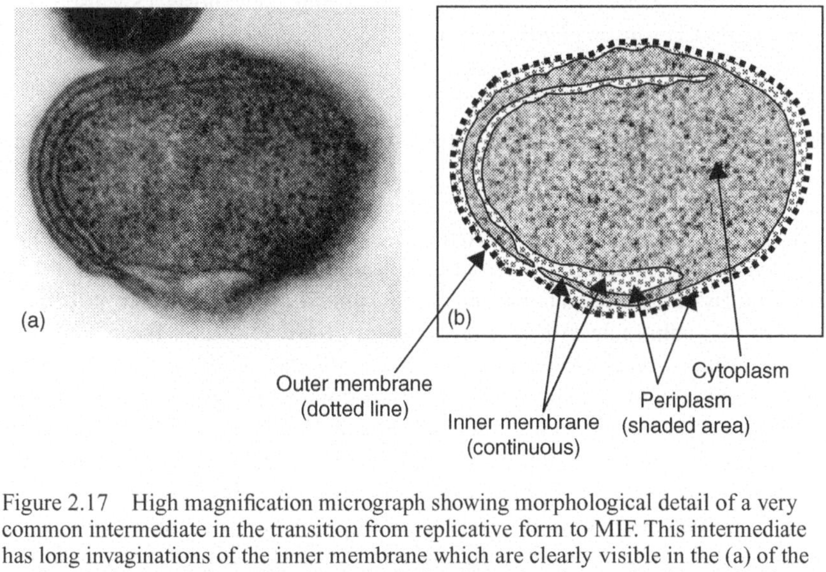

Figure 2.17 High magnification micrograph showing morphological detail of a very common intermediate in the transition from replicative form to MIF. This intermediate has long invaginations of the inner membrane which are clearly visible in the (a) of the section. An interpretation of the image is depicted as a drawing on the (b) panel, where the membranes, cytoplasm and periplasm have been labeled. For size reference, the thickness of the inner membrane is 5 nm. Micrograph by Gary Faulkner and Rafael Garduno, Dalhousie University, Halifax, Nova Scotia, Canada.

into simple nutrients. Thusly, the host phagocyte consumes portions of itself for survival during times of nutrient deprivation. *Legionellae* exploit this process to support rapid intracellular growth in the host phagocyte. But nutrients become limiting to the pathogen eventually within hours, from this process as the host cell is consumed. Nutrient depletion in the corrupted phagosome leads to the next phase in the pathogenic life cycle of *Legionella*.

The next phase is characterized by the response of exponentially growing *Legionella* to amino acid starvation as the nutrient content of the host is spent. The depletion of amino acids available to *Legionella* causes accumulation of uncharged tRNA inside the rapidly growing bacterial cells. This condition activates the ribosome-associated enzyme RelA. The RelA regulator converts guanosine triphosphate (GTP) to guanosine 3′,5′-*bis*-pyrophosphate (ppGpp) which positively activates the stationary phase sigma factor, σ, RpoS (Hammer and Swanson 1999). This cascade of events triggers virulence traits in *Legionella* including protist cytotoxicity factors and synthesis of bacterial flagella. A protein, IcmT, is necessary for pore-forming activity in the host cell membrane that leads to death of the host (Molmeret *et al.* 2002; Alli *et al.* 2003). Other cytotoxins are also apparently necessary including legiolysin (encoded by *lly*) and another cytotoxic protein called Msp which is a zinc metalloprotease.

2.3.3 Lysis

As *Legionella* becomes vigorously motile within massive intracellular vesicles (Figures 2.18 and 2.19) the spent host becomes completely overwhelmed by the infection and eventually it bursts to release motile and virulent progeny that seek opportunities to repeat this life cycle by infecting new hosts (Swanson and Hammer 2000; Swanson and Bachman 2002).

2.3.4 Ecological significance

The legionellae are natural parasites of aquatic and soil protists and incidental parasites of mammalian phagocytes. The ecological result of protist cell death is to release into the aquatic environment many hundreds or thousands of motile virulent bacteria in search of new hosts to infect. In low nutrient aquatic environments, *Legionella* loses the ability to replicate but not the ability to infect new hosts. After 125 days in sterilized tap water, *L. pneumophila* JR32 cells were still *v*iable *b*ut were *n*on-*c*ulturable (VBNC) on standard growth medium; these survivors were resuscitated when *Acanthamoeba castellanii* was added to the microcosm and were still infective (Steinert *et al.* 1997). In contrast to this, infection of mammalian host cells appears to be incidental (or accidental); the virulent progeny of these infections ultimately can benefit only in an ecologically insignificant way, apparently, since person-to-person transmission has never been observed.

An important aspect of this host–parasite association is that *Legionellae* are significantly more resistant to thermal and chemical disinfection when associated with their *Acanthamoebae* host (Storey 2004a) and more resistant to chemical disinfection when associated with *Hartmannella* in biofilms (Donlan 2005).

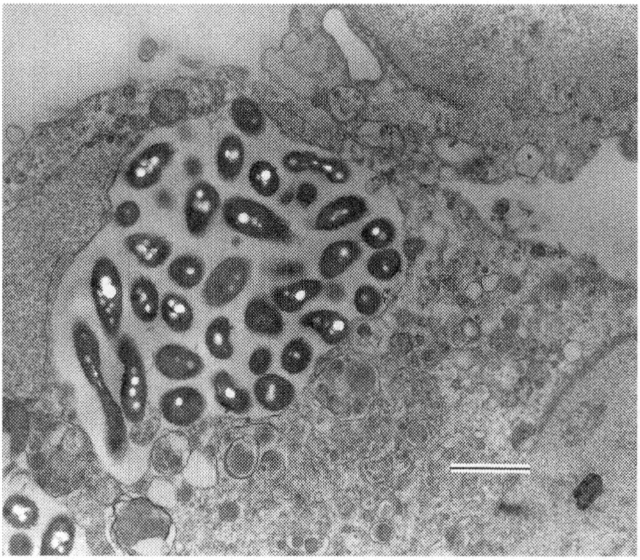

Figure 2.18 Late phase of infection in a HeLa cell that shows the unequivocal signs of deterioration: electron translucent cytoplasm and vesiculation. At this point most of the parasitic *Legionella* inside the dying host cell have differentiated into MIFs as shown in this micrograph. MIFs have a very dark (electron dense) cytoplasm, a complex often multilayered cell wall and contain numerous inclusions that appear as bubbles in this micrograph. Bar = 1 μm. Micrograph by Gary Faulkner and Rafael Garduno, Dalhousie University, Halifax, Nova Scotia, Canada.

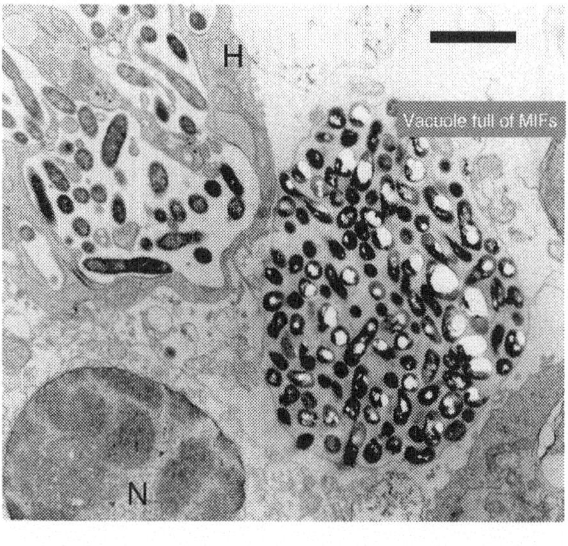

Figure 2.19 Replicative *L. pneumophila* in HeLa cells (H: on the top left corner) and a dying Hela cell (N: the nucleus of the dying HeLa cell) releasing a vacuole containing hundreds of MIFs. In a side-by-side picture like this, it is easy to spot the morphological differences between replicative forms and MIFs. Bar = 2 μm. Micrograph by Gary Faulkner and Rafael Garduno, Dalhousie University, Halifax, Nova Scotia, Canada.

2.4 MOTIVE TO CONTROL PROTIST HOSTS

Further motive to abate pathogen hosts in water systems comes from evidence that at least one pathogenic yeast-like fungus (*Cryptococcus neoformans*) can cause meningitis in immunocompromised patients with increased virulence as a result of interactions with aquatic amoeba (Steenbergen 2001). Still another potentially important finding is that *Vibrio cholerae* (the causative agent of cholera) survived its association with freshwater amoeba and proliferated in laboratory microcosms much more effectively in the presence of amoeba and could be re-isolated from encysting *Naegleria grubei* (Thom *et al.* 1992). In fact, coliforms and other bacterial pathogens survived within protozoa during chlorination; chlorine efficacy against bacteria significantly decreased for protozoan-ingested *Eschericia*, *Citrobacter*, *Enterobacter*, *Klebsiella*, *Sallmonella*, *Shigella*, *Legionella*, and *Campylobacter*; they all survived ingestion by protozoa; free chlorine was at least 50 times less effective against bacterial pathogens ingested by the ciliated protozoan *Tetrahymena*; bacterial resistance to digestion by predatory protozoans was proposed as a precursor of pathogenicity and it may be a mechanism for survival in aquatic environments (King *et al.* 1988). Protozoa grazing upon biofilms are reservoirs of pathogenic bacteria (Brown and Barker 1999).

Due to the similarities between phagocytic protists and human phagocytes, another motive to control exposure of susceptible humans to freshwater amoeba from engineered water systems is to protect against the emergence of new communicable diseases. Even after many hundreds of thousands of legionellosis cases worldwide since emergence of the disease in the 20th century, there has never been a single documented case of person-to-person transmission of legionellosis (no secondary cases). This indicates that the pathogen has not yet adapted to sustainable life within the human respiratory system and confirms that the natural host for the parasite could not be alveolar macrophage. Eventually however, after enough exposure to susceptible human immune systems, *Legionella* or other parasites of freshwater amoeba (see Table 2.1) could sufficiently adapt to achieve communicable transmission such as occurs person-to-person by *Mycobacterium tuberculosis*, the causative agent of tuberculosis.

Free-living amoebae appear to proliferate readily in domestic water such as hot water systems in hospitals (Rohr *et al.* 1998). Amoebae were observed in 52% (29 of 56) of samples from six hospital hot water systems and 47% (23 of 49) of samples from moist areas in the same hospitals. *Hartmannella*, *Echinamoeba*, *Saccamoebae*, *Vahlkampfia* (in hot water), and *Acanthamoeba*, *Naegleria*, *Vanella* (in moist areas) were identified in this study. Further to their role as pathogen hosts, *Acanthamoeba* themselves are causative agents in potentially serious diseases, such as granulomatous amebic encephalitis, amebic keratitis (infections of the eye), and sinusitis (Marciano-Cabral and Cabral 2003).

3

Effect

Legionellae are causative in severe pulmonary disease that can cause permanent damage. Debilitating effects to kidneys and in the nervous system are common in advanced disease. Legionellosis can also cause much milder flu-like disease.

The World Health Organization classifies legionellosis as a serious and preventable disease (WHO, *Legionella* and the Prevention of Legionellosis, *in preparation* to publish in 2005, www.who.int/water_sanitation_health). A great lot of legionellosis is preventable because proliferation and transmission of the pathogen are the result of correctable conditions in engineered water systems that are in close proximity to susceptible populations. Legionellosis disease is serious because it can result in severe acute pneumonia and, even in patients who recover, it can cause long-term debilitating health effects.

3.1 PATHOGENICITY

Legionellosis is a respiratory disease. Inhalation (aspiration) of infective particles or contaminated water into the alveoli can lead to mild flu-like symptoms or severe pulmonary disease. The size of aerosol droplets or particles is relevant because diameters larger than about 8 µm are excluded from alveoli by normal terminal bronchiole. However, aspiration of water into the lungs is common and does not depend on droplet size of aerosol.

In severe disease, damage to the lungs occurs as the result of bacterial growth inside alveolar macrophage (white blood cells, WBCs) and type 1 alveolar epithelium

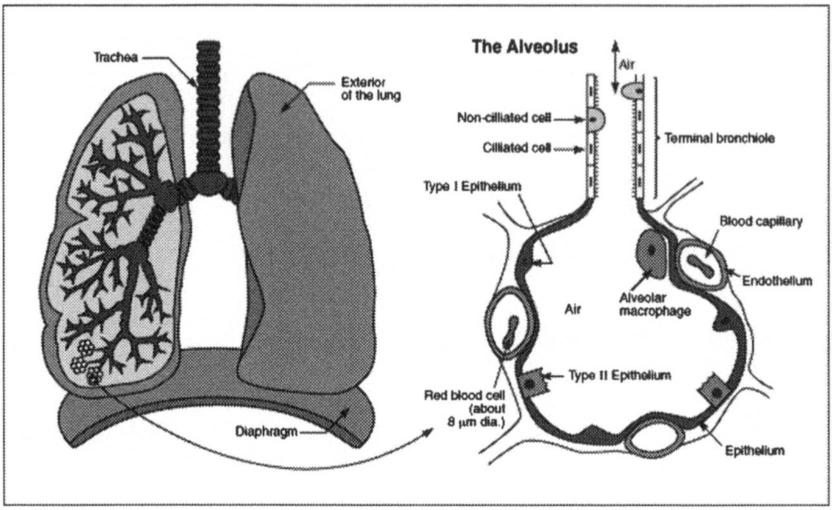

Figure 3.1 Pathogenicity depends on deep aspiration (inhalation) of *Legionella*-contaminated water, aerosolized droplets, or infective particles through the trachea, the bronchial passages, the terminal bronchiole, and finally into alveoli, the site of gas exchange between air and blood during respiration. The delicate structure of the alveolus is protected from particles larger than about 8 μm in diameter by ciliated cells in the terminal bronchiole. Particles less than this diameter or water may enter into the alveolar sac. In healthy individuals, phagocytic cells in the alveolus eliminate microorganisms and viruses that make it through all of the other defences in the lung. Alveolar macrophage and type I epithelium are susceptible to infections by *Legionellae*. (Adapted from *Microbial Waterborne Pathogens*, E. Cloete *et al.* (eds) ISBN: 1843390558; IWA Publishing 2004.)

cells. Figure 3.1 provides an anatomical sketch of the primary infection site for *Legionella*.

The alveolar epithelium consists of two cell types: alveolar types I and II cells. Although both cells are found in approximately equal numbers, type II cells are typically found in the corners of the alveoli, whereas type I cells are flattened and cover 90% of the alveolar surface area. Type II epithelium has been associated with much of the physiology of lung epithelium (e.g. surfactant secretion, salt movement, and progenitors for new epithelium). Type I cells facilitate gas exchange between air from the atmosphere and blood cells in capillaries. They also provide a protective barrier to blood capillaries and are endocytic. However, the primary protective cells in the alveolar sac, the alveolar macrophage (WBCs), are phagocytic. These cells and type I alveolar epithelium are susceptible to infection and intracellular growth by the *Legionellae* (Abu Kwaik *et al.* 1998).

3.1.1 Infectivity factors

Infectivity factors in *Legionellae* that lead to disease appear to be associated mostly with *dot/icm* genes that are involved in assemblage of the type IV protein secretion system; these 24 genes are central to pathogenesis. Biphasic killing of mammalian cells begins with apoptosis (programmed cell death). Induction of necrosis and lysis of the host then ensues. A protein, IcmT, specified by a gene in the *dot/icm* gene complex is essential for pore-forming activity and release of bacteria from the infected host cell. Pore-forming activity is important in pulmonary cytotoxicity and inflammation (Molmeret *et al.* 2002; Alli *et al.* 2003).

An interesting hypothesis is that phospholipase activity from the type II protein secretion in *Legionella pneumophila* (see Section 2.1.2) might cause damage to alveolar epithelial cells during infections of the lung by *Legionella.* Phospholipase from *L. pneumophila* degrades phosphatidylglycerol and phosphatidylcholine, the major constituents of lung surfactant, to the hemolytic agent lysophosphatidylcholine, which can induce acute respiratory distress syndrome and elicit inflammation (Flieger *et al.* 2002).

Free-living amoebae appear to be very important in legionellosis disease. Using laboratory animals (A/J and BALB/c mice) to model legionellosis disease, inhaled *Legionella*-infected *Hartmannella vermiformis* were significantly infective particles (Brieland *et al.* 1996, 1997a). Inhaled amoebae enhanced intrapulmonary growth of *Legionella* by providing a niche for bacterial replication (Brieland *et al.* 1997b). Two *Acanthamoeba* species copiously expelled 2–6 μm diameter membrane-bound protozoan vesicles laden with virulent bacteria; the amoebae remained viable, they produced a continuous source of *Legionellae*-filled vesicles by shedding dozens of potentially infective particles per amoeba in a 24 h period depending on conditions (Berk *et al.* 1998).

It is now clear that virulence of pathogenic *Legionellae* is enhanced by proliferation inside their protozoan hosts (Abu Kwaik *et al.* 1998). Thus the natural host–parasite relationship appears to be directly causative in the disease. Person-to-person transmission of the pathogen is not known to occur and there is only sketchy evidence of causes involving exposure other than from deeply aspirated aerosols contaminated with virulent *Legionella*.

The infective dose of *Legionella* is not known and of course, it varies depending on the susceptibility of the host. For this reason, it is not possible to set a scientifically defensible "safe" concentration of *Legionella.* Nor is it therefore, scientifically rational to set action (or "trigger") levels based on detected concentrations of *Legionella* in environmental samples (further discussion in Chapter 8, Validation and Verification).

In the early days of legionellosis diagnosis and treatment, mortality rates were very high approaching 20% and sometimes greater than 50% in immunosuppressed patients. Treatments with macrolide or quinolone antimicrobial agents (Section 3.2) that concentrate within macrophage cellular vesicles such as lysosomes have resulted in much lower mortality rates (Edelstein 2002). But it has now become clear that the severity of the disease can cause irreparable damage to the lung. Figure 3.2 shows chest X-ray analyses of a legionellosis patient during clinical presentation of symptoms and then after recovery. These radiographic images reveal significant and permanent damage to the lungs (Bartlett *et al.* 1986). Histology preparations of healthy and diseased lung tissues (Figure 3.3) illustrate the pathological effects of legionellosis.

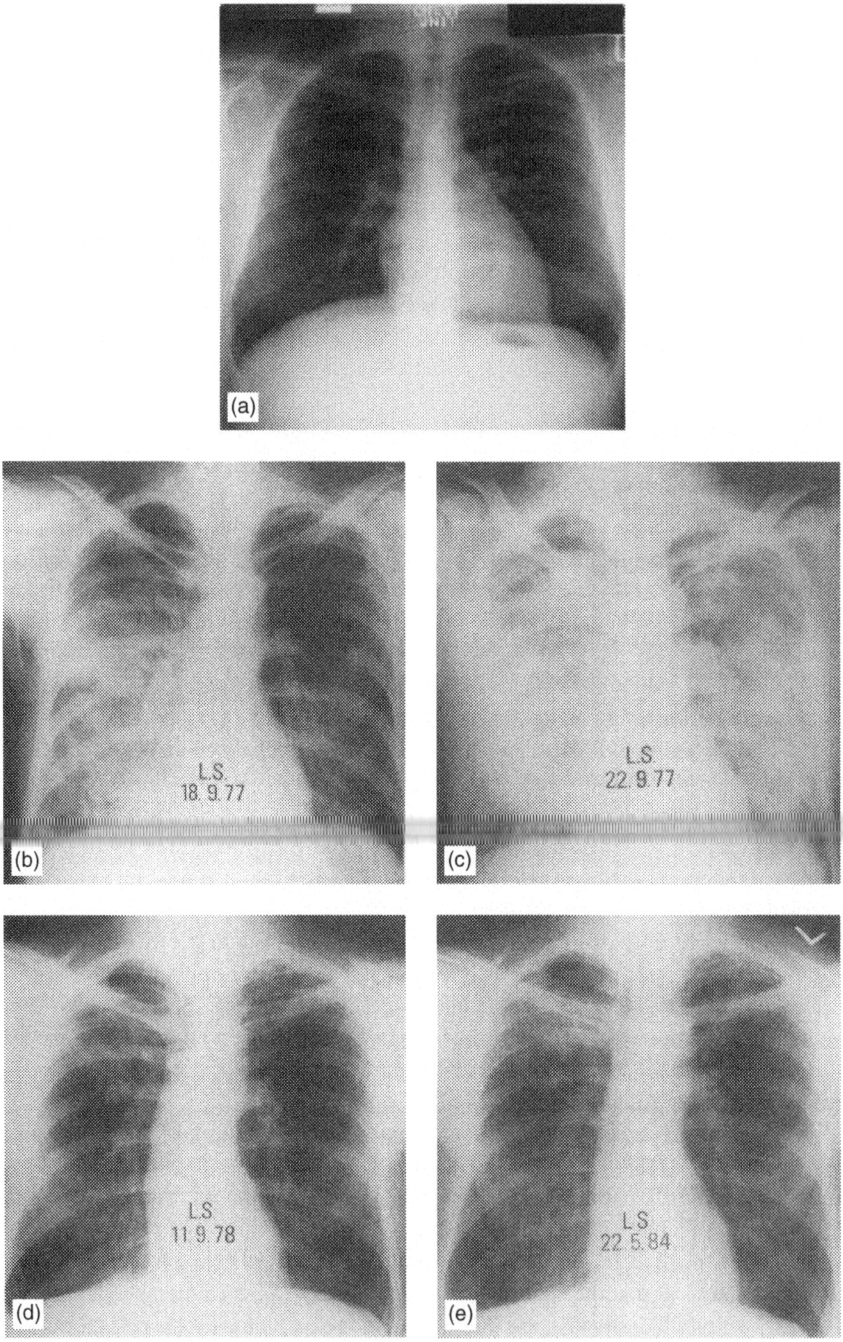

Figure 3.2 Chest X-rays of a normal healthy individual (a) and a legionellosis patient on admission (b), day 4 (c), 1 year post-infection (d), and 7 years post-infection (e). Long-term effects can be seen in the post-infection radiographic images. Dr. Tom Makin, Directorate Manager, Department of Medical Microbiology, The Royal Liverpool and Broadgreen University Hospitals, is gratefully acknowledged for his interpretation of these radiographs.

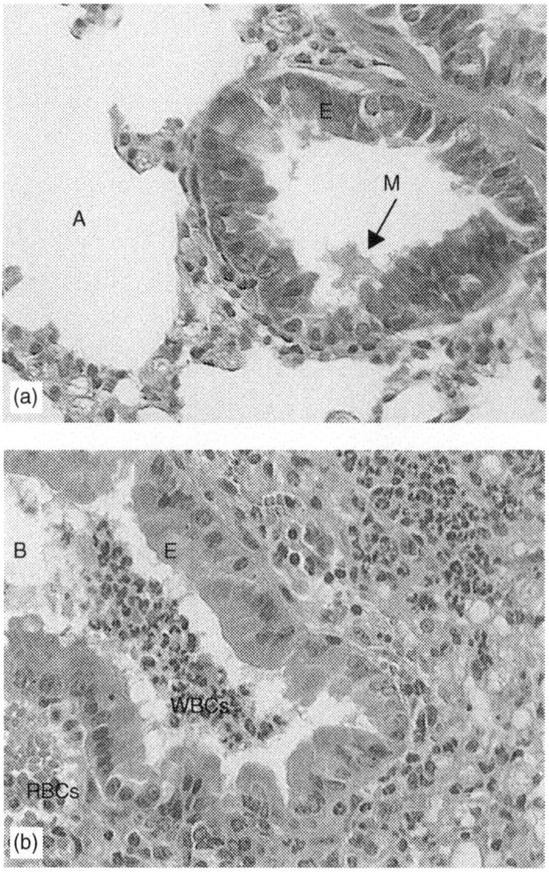

Figure 3.3 Hemotoxylin and eosin stains of normal mouse lung (a) and infected with *Legionella pneumophilia* SG1 (b) Visible are bronchiolar spaces (B), alveolar space (A, only visible in the healthy lung) surrounded by epithelium (E). An alveolar macrophage (M) is visible in the (a) healthy lung (M) bronchiolar space. In the infected lung, brochioles and peribroncholar regions are infiltrated with red blood cells (RBCs) and infected WBCs, which are not visible in large numbers throughout the normal lung. Micrographs by Dr. Jeffrey D. Cirillo, University of Nebraska, USA.

The infective process leading to legionellosis is summarized in Figure 3.4.

3.1.2 Incubation period summary

Legionellosis incubation period before onset of disease symptoms is widely regarded to be 2–10 days (less frequently, 14 days). However, a longer incubation period out to 19 or 20 days should be considered in outbreak investigations (the following data interpretation was by Thomas W. Armstrong, personal communication).

The premise for the widely cited 2–10 days incubation period for legionellosis appears to have started with the first report of the disease (Fraser *et al.* 1977).

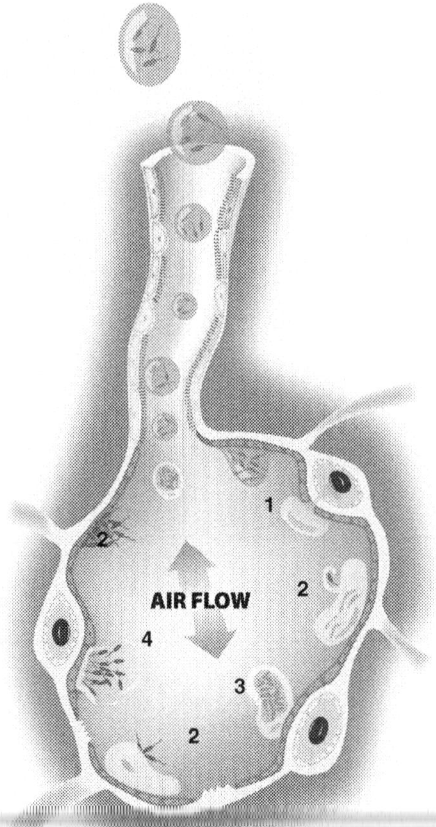

Figure 3.4 The infective process of *Legionella*. Pathogenic bacteria enter the alveolus via aspiration (inhalation) of water, infective particles, or microscopic droplets less than about 8 μm diameter. Macrophages (WBCs) enter the alveoloar space from blood capillaries (1) and become infected by *Legionella* through phagocytosis (2) *Legionella* grow intracellularly (3) in vesicles associated with the host cell endoplasmic reticulum. The host cell is overwhelmed by the parasite which has by then differentiated into motile cells (4) and is released as infective progeny in search of other cells to infect. Within 48 h, the alveolar space can become completely collapsed and filled with fluid (see Figure 3.3).

Most investigations use a case cut-off date of 10 (a few to 14) days following presumptive exposure. By definition therefore, these reports do not record cases with longer incubation periods.

The two primary data for determining the incubation period are: (a) definitely assigned date and source of exposure for cases and (b) an investigation exposure window longer than 10–14 days for case inclusion. Very few published legionellosis investigation reports are available to study incubation period. One notable study following a severe outbreak at a flower show in The Netherlands shows that 10 days is not an adequate cut-off for incubation (den Boer *et al.* 2002). Analysis of this outbreak data showed the following distribution for days to onset of symptoms (percentages are approximate):

- Onset within 3–10 days following exposure: 50% of cases
- Onset within 3–14 days following exposure: 90% of cases
- Onset over 14 days following exposure: 10% of cases
- Onset at days 18 and 19 following exposure: 1% of cases

The investigative team (den Boer *et al.* 2002) used an incubation period of 19 days and included reasonably certain data on exposure dates. An incubation period definition of 10 days maximum would have excluded approximately 50% of the cases from inclusion in that outbreak.

3.2 DIAGNOSIS AND THERAPY

Legionellosis may present in a variety of ways indicative of atypical community-acquired pneumonia (CAP) and produce clinical symptoms including hypoxia (blood oxygen below normal physiological levels), anorexia (loss of appetite), vomiting, non-productive cough, diarrhea, myalgia (muscle pain), headache, rising fever and chills, bradycardia (slowing of heart rate), and/or confusion/delirium. Radiographs are non-specific; most infiltrates are alveolar not interstitial. Blood work abnormalities include elevated liver enzymes, hyponatremia (low sodium), and hypophosphotemia (low phosphate). In clinic, extrapulmonary involvement (such as elevated liver enzymes) is characteristic for atypical CAP making legionellosis somewhat clinically distinct from typical CAP.

Bradycardia (slowing of the heart rate) is a cardinal feature of Legionnaires' disease; relative bradycardia clinically distinguishes legionellosis from atypical CAP caused by *Mycoplasma pneumoniae* (Cunha 2003). However, bradycardia does occur in atypical CAP caused by *Coxiella burnetti* and *Chlamydia pneumoniae*. It is interesting to note that *Coxiella* is phylogenetically closely related to *Legionella* and both *C. burnetti* and *C. pneumoniae* are now known to be natural parasites of freshwater amoeba such as *Acanthamoeba* (see Table 2.1; Essig *et al.* 1997; Maurin and Raoult 1999).

There are several methods for legionellosis diagnosis. Isolation of bacterial cultures from lung sputum samples is the preferred method because speciation and typing of the isolate can give important information (Chapter 8). Urinary antigen testing has become widely used in the clinic even though it is not very sensitive to *Legionellae* other than *L. pneumophila* serogroup 1 (Lp SG1) (Benson *et al.* 2000; Hayden *et al.* 2001). Polymerase chain reaction (PCR) detection and other molecular methods show promise for clinical diagnosis of legionellosis (Chapter 8).

3.2.1 Case definition

There is no international case definition for legionellosis. The recommended presumptive case is presence of specific *Legionella* antigen in respiratory secretion or urine, direct fluorescent antibody staining of the organism in respiratory secretions or lung tissue, or at least a 4 times rise in specific serum antibody titer to

Legionella species. The recommended confirmed case is (one or more of the following) isolation of *Legionella* from respiratory secretions, lung tissue, pleural fluid, blood, or other normally sterile fluids, at least 4 times rise in specific serum antibody to Lp SG1 (WHO 1997). In New South Wales, Australia, the case definition for Legionnaires' disease is a person with the signs and symptoms of pneumonia in whom *Legionella* has been isolated; or who has a positive *Legionella* antigen in their urine or respiratory secretions; or who has a fourfold or greater rise in antibody titer between acute and convalescent sera (McAnulty *et al.* 2001).

3.2.2 Antibiotic therapy

Several antibiotics are effective *in vivo* against *Legionella* especially those that concentrate intracellularly within lysosomes. The macrolide/azalide and quinolone drugs have the greatest efficacy. Erythromycin, the first macrolide drug ever commercialized (1950s), is very effective in treatment of legionellosis. Two newer macrolide antimicrobials, azithromycin, and clarithromycin, are now preferred. Quinolone drugs (especially ciprofloxacin and levofloxacin) are just as effective (Dedicoat and Venkatesan 1999). Empiric therapy for CAP (when the causative agent is not known) should be directed at typical and atypical pathogens. A respiratory quinolone or a cephalosporin (such as ceftriaxone) plus either a macrolide or a ketolide provides optimal coverage; doxycycline is especially useful because it covers the pathogens of both typical and atypical CAP (Cunha 2003).

Several treatment options for legionellosis are given in Table 3.1 (the table is adapted from Stout and Yu 1997; Dedicoat and Venkatesan 1999; Edelstein 2002;

Table 3.1 Preferred treatments for legionellosis.

Antibiotic	Dosage	Notes
Azithromycin	1000 mg once followed by 500 mg 2 times/day	5–10 days of treatment appears sufficient, IV or orally
Ciprofloxacin	400 mg IV or 750 mg orally 2 times/day	7–10 days
Levofloxacin	500 mg 1 times/day	Newer quinolones (gatifloxacin, moxifloxacin, etc.) also appear effective *in vitro*. IV or orally
Clarithromycin	500 mg 2 times/day	IV formulation not available in the USA
Erythromycin	1000 mg IV 4 times /day or 500 mg orally 4 times/day	Less often prescribed due to significant GI and ototoxicity (ear)
Doxycycline	100 mg 2 times/day	200 mg load is recommended
Tetracycline	500 mg 4 times/day	IV or orally
Rifampin	300–600 mg 2 times/day	Should only be used in combination with macrolide or quinolone in severely ill patients

IV: intravenous; GI: gastrointestinal.
Adapted from Edelstein 2002; Dedicoat and Venkatesan 1999; Stout and Yu 1997; and Srinivasan at www.hopkins-heic.org/infectious_diseases/legionella.htm.

and Srinivasan at www.hopkins-heic.org/infectious_diseases/legionella.htm; the dosages given are for patients with normal renal and hepatic function).

The newer macrolides and quinolones are considered drugs of choice. Tetracycline and doxycyline antibiotics have a broad spectrum of activity, are relatively safe, and can be recommended; they may even have antiprotozoal activity. Rifampicin is very active against extracellular and intracellular *Legionella* and is synergistic (*in vitro*) with erythromycin (Dedicoat and Venkatesan 1999). Typical duration of recommended treatments is 14 days in normal patients and 21 days in immunosuppressed patients.

In contrast, the β-lactam drugs (e.g. penicillin and ampicillin) do not penetrate the intracelluar compartments (phagosomes and lysosomes) in which *Legionella* grow and this may be why they are ineffective in treatment of legionellosis disease (Edelstein 2002). Further to resistance, many *Legionella* species (e.g. *L. pneumophila*) but not all species (e.g. *L. micdadei*) produce β-lactamase which inactivates penicillin (Marre *et al.* 1982). Although aminoglycoside antimicrobials penetrate lysosomal compartments, accumulation of the drug is very slow which is perhaps why they are ineffective in treatment of legionellosis (Butler *et al.* 1997).

In a retrospective review of the first recognized outbreak in Philadelphia in 1976, patients treated with erythromycin or tetracycline had a 50% lower mortality rate compared with patients treated with β-lactam drugs (Fraser *et al.* 1977).

3.2.3 Trends

Trends in empirical therapy for pneumonia and in new diagnostic tests appear to have had a major impact on clinical management of legionellosis (Benin *et al.* 2002). In the USA, analysis of 1980–1998 data from the passive surveillance system of Centers for Disease Control and Prevention (CDC) show that diagnosis by culture, direct fluorescent antibody and serological testing decreased significantly while diagnosis by urine antigen testing increased from 0% to 69% of all cases reported. This is significant because the urine antigen test can detect only Lp SG1 and therefore misses all other strains and species. Indeed, the frequency of isolates other than Lp SG1 decreased from 38% to 4%. The case-fatality rate significantly decreased from 34% to 12% for all cases and from 46% to 14% for nosocomial cases over the study period. The cause for this dramatic decrease in fatality rate is apparently due to confounding factors. More effective treatment appears to have reduced fatality from the acute disease whilst less culturing of clinical specimens appears to have resulted in fewer diagnoses of non-Lp SG1 infections and impeded outbreak investigations.

4

Extent

In regions with the best surveillance systems, more than 20 legionellosis cases per million population are officially reported each year. Even in these regions, the actual number of cases is believed to be more than they are reported.

So far, there is no evidence for person-to-person (secondary) transmission of the disease. Every case is therefore evidence that the victim was exposed to a pathogen-laden water system or soil.

Legionellosis is unique among other serious causes of pneumonia because its cause is transmitted to people from poorly maintained building water systems or contaminated soils.

4.1 INCIDENCE RATES

Data is now available to describe the extent of reported legionellosis disease. However, severe under-reporting of legionellosis is acknowledged. Surveillance data from Europe (Joseph 2002, 2004) and the USA (CDC 2003, 2004) indicates that there are on average about 6–10 cases reported per million population. Legionellosis is a notifiable disease in Australia. In New South Wales, Australia there were seven outbreaks of legionellosis between 1991 and 2000 but most cases (greater than 90%) were sporadic (not associated with an outbreak); the reported case rate was 9.3 per million population and the mortality rate was 10% (McAnulty *et al.* 2001).

In European countries with the best surveillance and mandated reporting requirements (e.g. Denmark, Netherlands, Belgium), incidence rates were more than 15 cases per million population in 1999 (Joseph 2002); reported case rates were similar in American States with specific guidance for legionellosis control and reporting protocols (e.g. the incidence rate in Pennsylvania and in Maryland was respectively, 21.4 and 14.6 cases per million population in 2004). There is very poor surveillance and/or no reporting requirement in many countries and in certain states of the USA. There may be more than 8000 cases (nearly 30 cases per million population) each year in the USA but extrapolations calculated from incidence rates in one state indicate that there may be up to 18,000 cases per year (Marston *et al.* 1994; Marston *et al.* 1997). Thus, it is not overstated to say that worldwide, hundreds of thousands of people are affected by the disease each year.

4.1.1 Demographics

The demographics of legionellosis are now well characterized. For example, during 2001 in the USA, 87% of all reported cases were age greater than 40 years; men were nearly twice as likely as women to be afflicted; children under 15 years rarely contract (less than 1% of reported cases) legionellosis (CDC 2003). These demographics are very similar in Australia (McAnulty *et al.* 2001) and in Europe. Many factors contribute to the susceptibility of disease including age, gender, immunosuppression, smoking, alcoholism, and autoimmune disease. Most legionellosis is associated with hot water domestic water service and industrial water systems like evaporative cooling towers. A significant travel-associated case rate is apparent in the data. Most reported legionellosis is sporadic community-acquired pneumonia (CAP) with less than about 5% reported from outbreaks of the disease (Marston *et al.* 1994).

4.1.2 Pneumonia

Pneumonia is the leading cause of death due to infectious diseases in the USA (Marston *et al.* 1997). There are more than 4 million episodes of CAP every year in the USA but only about 10% require hospitalization. The incidence rate of CAP requiring hospitalization in two Ohio counties in 1991 was 2668 per million population with an overall case-fatality rate of 8.8%. These results were used in a model to extrapolate that there are about 485,000 cases of CAP requiring hospitalization annually in the USA (Marston *et al.* 1997). About 15% of all CAP cases in the USA are caused by the "atypical pathogens" (*Legionella*, *Chlamydia*, *Coxiella*, and *Mycoplasma*) whereas 85% of CAP is caused by the "typical bacterial pathogens" (*Streptococcus*, *Haemophilus*, and *Moraxella*); in the USA, the causative agent of CAP is not determined in about 30% of cases (Cunha 2003) but some estimates of how often etiology of CAP goes undiagnosed is much higher (over 50%). Legionellosis may cause 4.1–20.1% of CAP cases requiring hospitalization and most legionellosis is associated with building water systems (reported in Rusin *et al.* 1997).

4.1.2.1 Outbreaks

Most legionellosis is sporadic and is associated with building water in close proximity to susceptible populations. However, explosive outbreaks of legionellosis result in far more public notice and governmental regulatory activity. Although relatively rare, outbreaks of legionellosis have been an important factor in public awareness from the beginning. Indeed, the Stafford, UK outbreak of 1985 lead to the first government-approved code of practice in 1987. Legionellosis pneumonia bears the name Legionnaires' Disease as a result of the first known outbreak in 1976 at the American Legion convention held in the Bellevue-Stratford hotel of Philadelphia; in this famous outbreak, 221 persons were afflicted and 34 (15%) people died.

The most spectacular outbreak of legionellosis occurred in July 2001 at Murcia, Spain (García-Fulgueiras *et al.* 2003). The source of the outbreak was proven to be from improperly controlled contaminated hospital cooling towers in the north of the city; there were 800 suspected cases, 449 confirmed. Hospitalization was necessary for 64% of the cases. There were six deaths directly related to this outbreak (less than 2% of confirmed cases). The remarkably low-fatality rate was due to quick detection of the outbreak, early diagnosis of the disease, and appropriate treatment of patients. The explosive qualities of the outbreak prompted clinicians to diagnosis accurately and initiate treatments immediately. This outbreak precipitated the first regulatory document with specific legal status in Spain (Spain 2001; Table 5.1) and it appeared just 20 days after onset of the outbreak.

Previously, in Victoria, Australia an outbreak also resulted in quick publication of government-approved code of practice supported by enacted legislated penalties for non-compliance after 93 confirmed cases and 3 deaths occurred in April 2000. In this outbreak, hundreds of people were exposed to *Legionella*-laden mist from an improperly treated cooling tower at the newly opened Melbourne Aquarium.

Prior to the Murcia, Spain outbreak the largest and most significant was that which occurred in The Netherlands at the Westfriese floral show of February 1999. One hundred eighty eight people who visited the fair were afflicted with legionellosis pneumonia (133 confirmed cases, 55 probable), 163 (87%) were hospitalized, 24 (21%) required mechanical ventilation, and 21 (11%) people died. The source of the outbreak was a contaminated whirlpool spa on exhibition at the flower show; another spa and a sprinkler system were also shown to be contaminated but in a careful epidemiologic study, the probable source of the outbreak was the new spa (den Boer *et al.* 2002).

In France, a highly publicized 2003 legionellosis outbreak at Lens, in Pas-de-Calais (in the north of France) afflicted 69 people of whom 9 victims died. A Noroxo petrochemical plant (part of the Exxon group) was identified as the source (ProMed news posting Tuesday, 13 Jan 2004).

4.1.2.2 Officially reported cases of legionellosis

Reported cases of legionellosis in the USA during 2004 are given by state in Figure 4.1. There is an uneven distribution of reported cases.

Many contributing factors result in uneven incidence reporting of legionellosis. When an individual patient seeks care, the clinician first decides whether to collect a diagnostic specimen. Currently available antibiotics are so effective against so many causes of pneumonia, that there is now less incentive to collect a

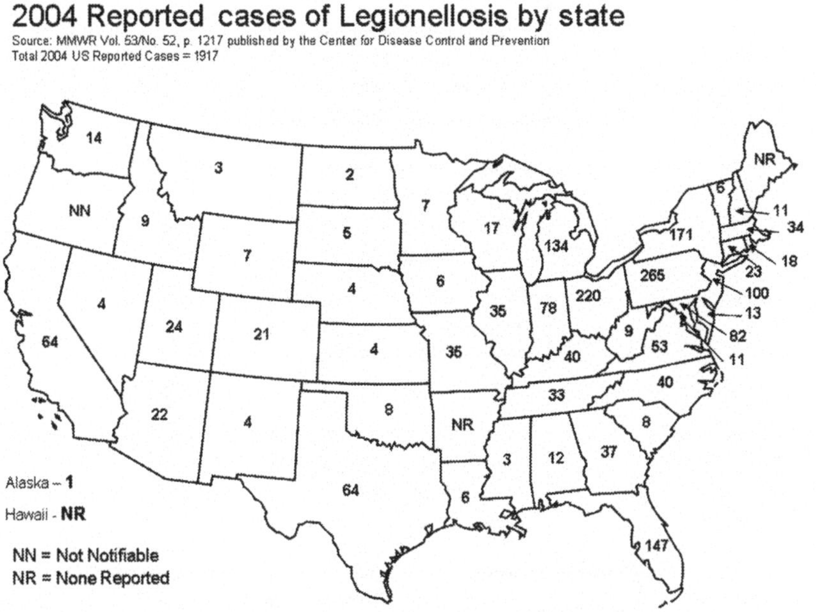

Figure 4.1 Reported cases of legionellosis in the USA show the results of very uneven surveillance (graphic by Larry Gess, Phigenics, LLC).

specimen and more of a trend toward empiric treatment. Patterns of practice vary widely from state to state. If a diagnostic specimen is obtained, it has to be processed properly. If a positive result is obtained, then it has to be reported to the local Health Department then to the state Health Department and then to the Centers for Disease Control and Prevention (CDC). There is a significant and highly variable lag in reporting from the states to the CDC. Figure 4.2 illustrates the sequence of events of the process to report positive results to the CDC.

It is useful to look closely at the officially reported incidence rate data expressed as reported cases per million population (Tables 4.1–4.4). Variations in population demographics may account for some of the incidence rate differences between states. However, grouping the states by decreasing incidence rate shows no clear pattern by demographic which suggests that differences are due to uneven reporting procedures.

The data in Table 4.1 show that about 20% of the US population (59.8/293.7) live in states for which reported incidence rates range from 10 to 21 cases per million population. Of these, only Pennsylvania and Maryland have published guidance documents for legionellosis control and prevention. Taking Pennsylvania as the most progressive and accurate reporter of legionellosis cases in the USA, an extrapolation of their case rate to the entire US population suggests that more than 6000 cases (293.7 × 0.214) would be reported if every state reported to that extent.

The data in Table 4.2 show that about 22% of the US population (65.8/293.7) live in states for which reported incidence rates range from 5 to 9 cases per million

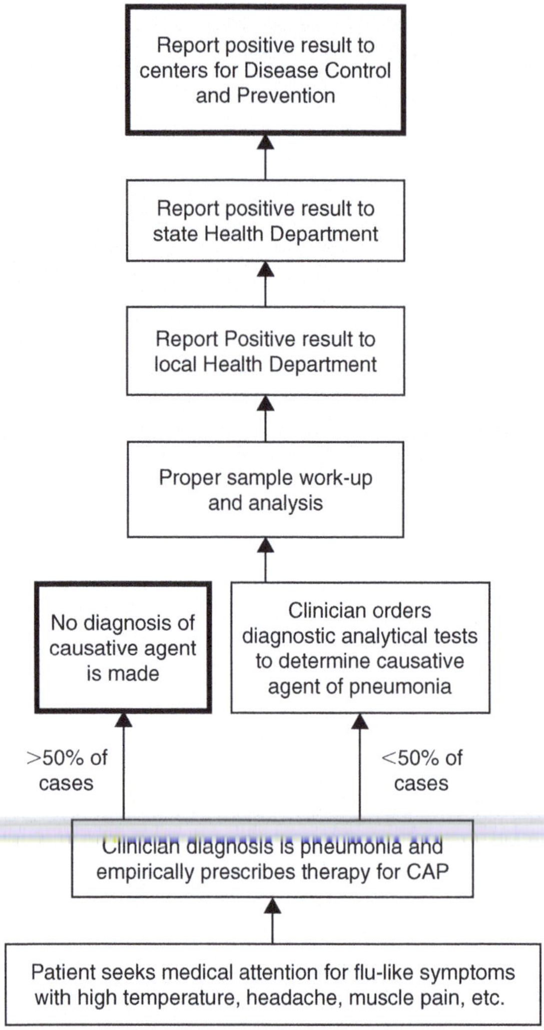

Figure 4.2 The process and sequence in the USA for reporting positive legionellosis results from clinical specimens to the CDC (derived from discussions with Dr. Matthew Moore, MD, Medical Epidemiologist, Centers for Disease Control and Prevention, Atlanta, GA).

population. Compared to Pennsylvania, these states appear to be under reporting legionellosis by over 60% (1–7.9/21.4).

The most striking result from looking at incidence rate data by state is to discover that 57% of the US population (167.9/293.7) lives in states with incidence rates reported as less than five cases per million population (Table 4.3). For instance, if the reported incidence rate in California was the same as in Pennsylvania, then there would be expected to have been 768 cases (35.9 × 21.4) of legionellosis in 2004, not just 65. In Illinois, there would have been 272 cases (12.7 × 21.4) instead of 35 reported cases. These data indicate that legionellosis is severely under-reported in the USA.

Table 4.1 US states (and the District of Columbia) with greater than 10 reported legionellosis cases/million population in 2004.

	Population (millions)	2004 Reported cases	Cases/million
United States	*293.7*	*1917*	*6.5*
Pennsylvania	12.40	265	21.4
DC	0.60	11	18.3
Ohio	11.50	220	19.1
Rhode Island	1.00	18	18.0
Delaware	0.80	13	16.3
Maryland	5.60	82	14.6
Wyoming	0.50	7	14.0
Michigan	10.10	134	13.3
Indiana	6.20	78	12.6
New Jersey	8.70	100	11.5
Utah	2.40	24	10.0
Subtotal	**59.8**	**952**	**15.9**

Data source: US census estimates April 1, 2004; CDC *MMWR*, week 53, 2004.

Table 4.2 US states with 10–5 reported legionellosis cases/million population in 2004.

	Population (millions)	2004 Reported cases	Cases/million
United States	*293.7*	*1917*	*6.5*
Vermont	0.6	6	10.0
Kentucky	4.1	40	9.8
New York	19.2	171	8.9
New Hampshire	1.3	11	8.5
Florida	17.4	147	8.4
Virginia	7.5	53	7.1
South Dakota	0.8	5	6.3
Idaho	1.4	9	6.4
Missouri	5.8	36	6.2
Tennessee	5.9	33	5.6
West Virginia	1.8	9	5.0
Subtotal	**65.8**	**520**	**7.9**

Data source: US census estimates April 1, 2004; CDC *MMWR*, week 53, 2004.

Analysis of the reported incidence rate data from Europe shows very much the same kind of variation. When the data are sorted by reporting country, they suggest that the range in incidence rate appears to be largely due to differences in surveillance requirements. However, the effect of very large outbreaks such as in Murcia, Spain in 2001 also heightened awareness and led to comparatively more proactive diagnosis in Spain. A rate of less than one case per million population was reported from the Czech Republic, Latvia, Northern Ireland and Poland in 2000, Ireland, Poland and the Slovak Republic in 2001 and Estonia, Poland and the Slovak Republic in 2002 (not listed in Table 4.4).

Table 4.3 US states with less than five reported legionellosis cases/million population in 2004.

	Population (millions)	2004 Reported cases	Cases/million
United States	*293.7*	*1917*	*6.5*
North Carolina	8.5	40	4.7
Colorado	4.6	21	4.6
Massachusetts	6.4	28	4.4
Connecticut	3.5	15	4.3
Georgia	8.8	37	4.2
Arizona	5.7	22	3.9
Montana	0.9	3	3.3
North Dakota	0.6	2	3.3
Wisconsin	5.6	17	3.0
Texas	22.5	64	2.8
Illinois	12.7	35	2.8
Alabama	4.5	12	2.7
Nebraska	1.7	4	2.4
Oklahoma	3.5	8	2.3
Washington	6.2	14	2.3
New Mexico	1.9	4	2.1
Iowa	3.0	6	2.0
South Carolina	4.2	8	1.9
California	35.9	65	1.8
Nevada	2.3	4	1.7
Alaska	0.7	1	1.4
Kansas	2.7	4	1.5
Minnesota	5.1	7	1.4
Louisiana	4.5	6	1.3
Mississippi	2.9	3	1.0
Arkansas	2.[illegible]	0	0.0
Hawaii	1.3	0	0.0
Maine	1.3	0	0.0
Oregon	3.6	0	0.0
Subtotal	**167.9**	**430**	**2.6**

Data source: US census estimates April 1, 2004; CDC *MMWR*, week 53, 2004.

Table 4.4 Selected European countries reporting through EWGLI in 2002 (Joseph 2004).

	Population (millions)	2002 LD cases	Cases/million
All reporting countries	*467*	*4696*	*10.1*
Spain	40.5	1380	34.1
Denmark	5.4	104	19.3
The Netherlands	16.1	288	17.9
France	60.2	1018	16.9
Italy	57.8	605	10.5
England and Wales	52.9	382	7.2
Germany	82.4	288	3.5
Subtotal	**315.3**	**4065**	**9.9**

Table 4.5 Occurrence of *Legionella* in engineered water systems.

Engineered water system	% Positive	Location/notes	Reference
Residential water supply			
Drinking water	3–33	USA	Rusin *et al.* 1997
Groundwater	56–83		Riffard *et al.* 2001
Water system biofilms	29		Riffard *et al.* 2005
Potable water taps (domestic and institutional)	61	96% were less than 1000 CFU/ml	Atlas 1999
Single-family residences	6	Canada	Marrie *et al.* 1994
Multi-family residences	25	Canada	Marrie *et al.* 1994
Water heaters	21–79	Europe	Tiefenbrunner *et al.* 1993
Showers, spas, and fountains			
Whirlpool spas	27	USA	Miller 2002
Decorative fountains	15–19	Singapore	Heng *et al.* 1997
Showers	8–92	Europe	Tiefenbrunner *et al.* 1993
Warm or hot water systems			
Cooling towers	51	USA	Miller and Kenepp 1993
Cooling towers	36	Singapore	Heng *et al.* 1997
Cooling towers	47	Finland	Kusnetsov *et al.* 1997
Cooling towers	90	Detection by PCR	Koide *et al.* 1993
Hospitals	71	USA (western PA)	Squier *et al.* 2005
Hospitals	83	USA	Goetz *et al.* 1998
Hospitals	47	France (Paris)	Nahapetian *et al.* 1991
Hospitals	68	SE Germany	Lück *et al.* 1993
Hospitals	70	Saxony	Habicht *et al.* 1988
Outpatient clinics	50	SE Germany	Lück *et al.* 1993
Dental offices	58	SE Germany	Lück *et al.* 1993
Public buildings	85	SE Germany	Lück *et al.* 1993
Hotels	18	Saxony	Habicht 1988
Private households	65	Mostly less than 100 CFU/ml	Lück 1993
Apartment buildings	30	Mostly in showers	Zacheus 1994

4.2 EPIDEMIOLOGY

4.2.1 Sources

Legionellae are common in freshwater aquatic environments and prevalent in engineered water systems. Survey data of such systems indicates close association of the opportunistic pathogen with humans in habitation (Table 4.5).

Legionella is commonly found in drinking water sources especially from groundwater and in biofilms (Riffard *et al.* 2005). Most legionellosis is associated with domestic water (potable water) service in buildings (Cordes *et al.* 1981; Shands *et al.* 1985; Schlech 1990; Lee *et al.* 2002; WHO 2002). However, utility water, such as cooling water systems, are also significantly associated with legionellosis (Dondero *et al.* 1980; Cordes *et al.* 1980; Garbe *et al.* 1985; Miller and Kenepp 1993; Keller *et al.* 1996).

Showers and whirlpool spas are believed to be among the most common sources of exposure to *Legionella*. In the UK, an epidemiologic study of 41 legionellosis

Table 4.6 Sources of reported legionellosis infections in Europe, 2000–2002 (Joseph 2004).

Source of infection	Reported cases 2000–2002	% Total
Community-acquired	3916	37.9
Unknown source	3399	32.9
Travel abroad	1424	13.8
Hospital-acquired	889	8.6
Travel home	687	6.7
Travel unknown	7	0.1
Total reported infections	**10,322**	**100**

cases in Nottingham between 1997 and 2000 showed that in 39% of cases, significant quantities of *Legionella* sp. were cultured from the domestic water service associated with the patients and there was no statistically significant correlation between proximity or exposure to cooling towers or other industrial sources of water for any of these patients (Lim *et al.* 2003). In the USA, families of 21 victims agreed to allow testing of their home water and *Legionella* was found in 24% of them; two of the patients studied died of the infections (Stout 2004, 2005).

Traces of *Legionellae* have been found in all phases of wastewater treatment (Palmer 1993, 1995). One species (*L. longbeachae*) is associated with potting soil mixes, aggressively infects *Tetrahymena* (a ciliated protozoan), and is known to have caused legionellosis in Australia and in Japan (Steele and McLennan 1996; Broadbent 2002).

Officially reported sources of infection in Europe show that about 9% of legionellosis is acquired in hospital (Table 4.6). This is particularly of concern since exposure to the pathogen is very dangerous for ill people in hospital.

Hospital-acquired and healthcare-related legionellosis rates are difficult to estimate because diagnosis is not usually made (see Section 4.1.2.2 and Figure 4.2). Assuming the nationwide incidence rate of legionellosis in the USA is that of Pennsylvania (21.4 cases per million population, Table 4.1) and assuming that 9% of all legionellosis cases are nosocomial, then in 2004 there must have been about 560 nosocomial cases in the USA.

But apparently the nosocomial legionellosis case rate is much higher than 9% of all legionellosis cases in the USA. In Pennsylvania, a survey of 48 hospitals revealed that 33% of all legionellosis cases were nosocomial and that 71% of hospitals water systems were colonized with *Legionella* bacteria prior to implementation of proactive guidance to control legionellosis (Squier *et al.* 2005). Extrapolating the data from Pennsylvania, which is the American state with the most active surveillance, to the rest of the country suggests that there may be nearly 2000 nosocomial legionellosis cases each year in America. The mortality rate for nosocomial legionellosis was 40% making it one among the most deadly *preventable* diseases known.

That legionellosis disease is in fact preventable (see Part 2, Prevention) has now been proven in Pennsylvania. After implementing proactive guidance for environmental surveillance of *Legionella* in healthcare water systems and disinfection of systems that were colonized with the pathogen, the nosocomial incidence rate of legionellosis declined from 33% to 9% in Pennsylvania (Squier *et al.* 2005).

PART TWO: Prevention

5

Guidance

Most victims of legionellosis were exposed to an improperly managed building water system. Therefore, most cases of legionellosis are preventable. An extensive body of published guidance on legionellosis prevention and control is now available worldwide.

5.1 CODES OF PRACTICE

Codes of practice, guidelines, standards, or best practices are the collaborative technical consensus opinions of experts from government, academe, and industry. These guidance documents define a standard of care against which the owner of an engineered water system could be judged in an accusation of negligence or breach of health and safety law. Thus, guidance publications are extremely important even in countries with no specific regulatory penalty or special legal status to enforce an approved code of practice (ACOP) to prevent legionellosis.

5.2 LAWS, REGULATIONS, AND STANDARDS

Codes of practice are not laws nor are they regulations. Rather, they give advice on how to comply with laws and regulations. The general legislative framework that applies to most government-approved guidance documents is given in Figure 5.1.

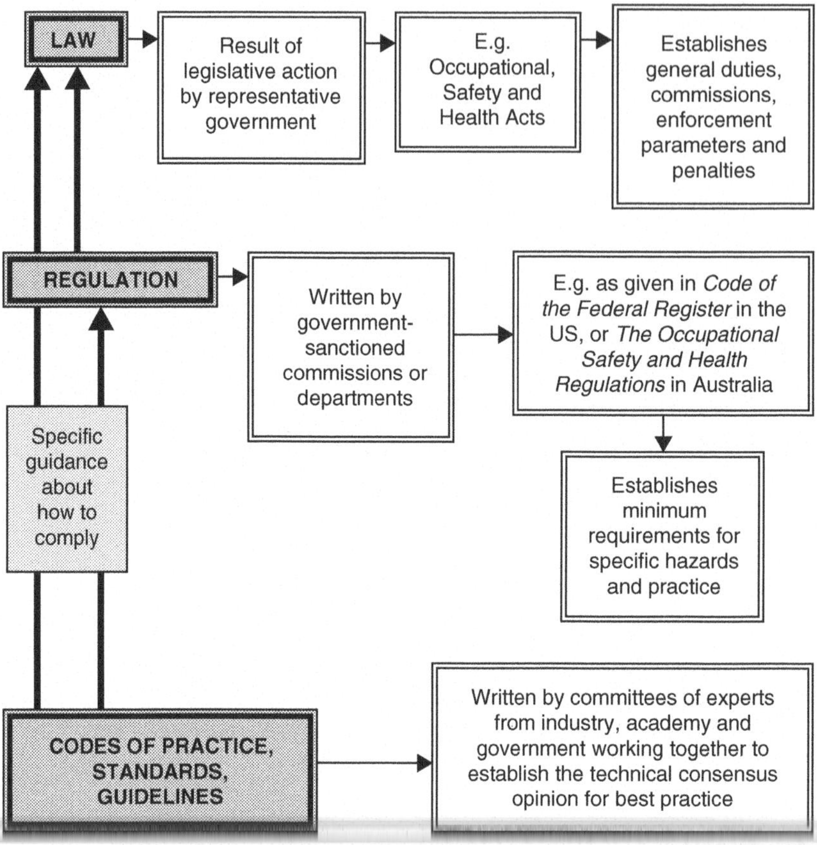

Figure 5.1 Codes of practice, standards and guidelines provide technical consensus opinions about how to comply with the established minimum requirements (regulations) and in the general duties established by legislated acts of government (laws).

5.2.1 Government-approved or sponsored codes of practice

Government-approved or sponsored guidance about how to comply with laws and regulations has been published for control of *Legionella* and preventing legionellosis. Table 5.1 lists the most well-known publications.

5.2.2 The United Kingdom

The most cited government-approved guidance for the control of *Legionella* is from the UK Health and Safety Commission (HSC 2000) in their *Approved Code of Practice and Guidance* (often referred to in the UK as "L8" or the "ACOP"). It is advice on how to comply with the *Control of Substances Hazardous to Health Regulations* (1999, regulation 6–9 and 12) and the *UK Health and Safety at Work Act* (1974, Sections 2, 3, 4, and 6). "L8" is the result of guidance revisions (EH48 → HS(G)70 → L8) beginning in 1987 following the Stafford, UK outbreak of 1985.

Table 5.1 Government-approved or sponsored guidance for prevention of legionellosis.

Country	Year	Title	Reference
France	2005	Various documents describing new regulations and guidance for preventing legionellosis and requiring use of hazard analysis and control	France 2005a, b, c
USA	2004 2003	Guidelines for environmental infection control in healthcare facilities: recommendations of CDC and the Healthcare Infection Control Practices Advisory Committee (HICPAC)	CDC 2003; 2004
Australia/ New Zealand	2002	AU/NZ Standards: 3666.1, 3666.2; Air-handling and water systems – Microbial control	Australia 2002
Victoria, Australia	2002	Code of practice for water treatment service providers	DHS 2001, 2002
Singapore	2001	Code of practice for the control of *Legionella* bacteria	Singapore 2001
France	2001	Sanitary and hygiene criteria for preventing and controlling legionellosis	France 2001
Spain	2001	Guide to good practice: *Legionella* in cooling towers	Spain 2001
USA	2001	*Legionella*: Drinking Water Health Advisory	EPA 2001
USA	2001	Environmental infection control in healthcare facilities	CDC 2001
Western Australia	2000	Code of Practice – Prevention and Control of Legionnaires' Disease	Western Australia 2000
Hong Kong	2000	Prevention of Legionnaires' Disease – Code of Practice	Hong Kong 2000
USA (Maryland)	2000	Report of the Maryland Scientific Working Group to study *Legionella* in water systems in Healthcare Institutions	Maryland 2000
UK	2000	Legionnaires' Disease: the control of legionella bacteria in water systems, ACOP and guidance (often referred to as "L8")	HSC 2000
Australia/ New Zealand	2000	AU/NZ standard: performance-based maintenance of cooling water systems	Australia 2000
The Netherlands	2000	The temporary mains water (*Legionella* prevention) regulation	Netherlands 2000
USA	2000	Guideline: minimizing the risk of legionellosis associated with building water systems	ASHRAE 2000
USA (Texas)	1999	Report of the Texas Legionnaires' Disease Task Force	Texas 1999 Rev 2004
USA	1998	Technical manual, Chapter 7: Legionnaires' disease	OSHA 1998
USA (PA)	1997	Approaches to the prevention and control of *Legionella* infection	PA 1997
Germany	1997	Technical improvements for the reduction of the growth of *Legionella*	DIN 1997; DVGW 1996
Australia/ New Zealand	1995	AU/NZ standard: operation and maintenance; design, installation and commissioning	Australia 1995a, b
UK	1991 Rev 1995	The control of legionellosis, ACOP	HS(G) 70 HSC 1995
UK	1987	Legionnaires' disease	EH48 1987

The document is in two parts: The actual code of practice and, in the second part, specific guidance on how to comply with the code of practice. It is in this second part, published in November of 2000, that advice on pathogen monitoring and action levels appear for the first time in UK guidance. An industry-sponsored guide to implementing the ACOP is helpful (CIBSE 2002).

The UK ACOP has become the basis for many other guidance documents around the world. For example, the Codes of Practice approved in Singapore, Hong Kong, and Japan have been principally derived from the UK ACOP.

The ACOP applies to the control of *Legionella* bacteria in any undertaking involving a work activity and to premises in connection with a trade or business where water is used or stored and where there is a means of creating and transmitting water droplets which may be inhaled, thereby causing a reasonably foreseeable risk of exposure to the pathogen. Such water systems from which reasonably foreseeable risks of exposure to the pathogen exist are cooling towers, evaporative condensers, hot and cold domestic (potable) water systems, and other systems containing water warmer than 20°C which may release sprays or aerosols during operation and/or maintenance.

The UK ACOP requires risk assessment driven approach to risk management. Specifications and detailed parameters for performing risk assessments are given. Risk management procedure and guidance is given for implementing precautions to reduce the identified risks.

For perspective on how risk assessment driven plans compare to hazard analysis and control plans, see Section 6.2.3.1 and in Section 6.3.4.

5.2.3 Australia/New Zealand and Asia Pacific

The Australian/New Zealand Standard was published first in 1995, overhauled in 2000 with Parts 1 and 2 revised in 2002. It gives specifications on how to comply with public health legislation in Australia and New Zealand. For instance, in New South Wales, the Public Health Act of 1991 is the law for which the Public Health (Microbial Control) Regulation 2000 (amended July 2003) gives guidance. These regulations are based upon the AU/NZ Standard (AS/NZS 3666.1: 2002 Air-handling and water systems of buildings – Microbial control – Design, installation and commissioning; 3666.2: 2002 Air-handling and water systems of buildings – Microbial control – operations and maintenance; 3666.3: 2000 Air-handling and water systems of buildings – Microbial control – Performance-based maintenance of cooling water systems).

The Code of Practice in Victoria, Australia (DHS 2002) is crafted for emphasis on developing a site-specific risk management plan. This guidance was the rapid direct result of a legionellosis outbreak in which several government officials, including the Australian Finance Minister, were confirmed as having contracted legionellosis during a reception at the Melbourne Aquarium in April 2000. Hundreds of other notable political figures including the Prime Minister were put at risk and the opposition leader underwent tests for the bacteria after experiencing symptoms. A total of 93 people were confirmed to have contracted the disease, 7 were hospitalized in critical condition, and 3 people died.

Today in Victoria, Owners of land upon which a cooling tower system is situated are required to: (1) register the system, (2) develop, implement, and review a risk management plan, (3) implement an improved maintenance program and (4) keep records of maintenance and test results for audit/inspection purposes. Specific legislation to enforce this code of practice was established by the government. The maximum penalty for failing to complete a risk management plan in Victoria, Australia is AU$6000 per cooling water system. The maximum penalty for failing to register a cooling tower system is AU$12,000 (DHS 2001a, b, 2002).

In Western Australia, the Minister for Labour Relations approved a code of practice that applies to all workplaces covered by the Occupational Safety and Health Act of 1984 (Western Australia 2000).

5.2.4 Europe

Taking regulatory requirements in the Netherlands as an example (Netherlands 2000), Owners of collective water installations, for example, in hospitals, hotels, campsites, and swimming pools, are obliged to conduct a risk analysis and if necessary take measures to prevent health risks posed by *Legionellae*. Actions to correct deficiencies and reduce risk must be specified in a control (risk management) plan.

The Spanish Ministry of Health and Consumer Affairs published a Royal Decree in July 2001 that establishes sanitary and hygiene criteria for preventing and controlling legionellosis. Not coincidentally, the Spanish legionellosis guidance and regulation was published immediately after the Murcia outbreak which to this date is the largest to have occurred anywhere in the world (see Section 4.1.2.1). The Spanish guidance is generally consistent with guidance given elsewhere. For example, cleaning and disinfection is required for cooling towers but it should only be done after risk characterizations of the system and the efficacy of the treatments should be shown (validated) by pathogen analytical testing before and after. Thus, hazard control activities should be done within the context of an established plan that includes properly defined risk management practice.

5.2.4.1 France

In France, the regulatory situation has rapidly evolved due to high nosocomial incidence rates especially in Paris and a highly publicized 2003 outbreak at Lens, in Pas-de-Calais (in the north of France). Of the 69 people who contracted legionellosis in this recent outbreak, 9 victims died. A Noroxo petrochemical plant (part of the Exxon group) was identified as the source (ProMed news posting Tuesday, January 13, 2004).

The legionellosis control regulations in France were upgraded in December, 2004 and officially approved by the government in May 2005 (France 2005a, b, c). They were upgraded from older regulations that had been previously consistent with guidance throughout Europe. The new regulations are much more specific and process-oriented. The new regulations for cooling tower water in France are the most extensive and specific in the world at the present time. They are to be enforced after July 1, 2005. Several highly significant activities are now required by law in France as follows.

Perhaps the most significant change in France is that legionellosis is to be prevented through *Legionella* hazard analysis and control. This process to manage risk is required for any water system in which cooling towers are used. Implementation of the hazard analysis and critical control point (HACCP) system (or any hazard analysis and control system that is functionally identical to HACCP) is now required by law in France. The HACCP system is not a risk assessment-driven process such as is required in the UK. For perspective about hazard analysis and control and how it compares to risk assessment-driven plans, see Chapter 6.

Another important change in the newly approved French regulation is that open cooling tower systems with thermal capacities greater than 2000 MW must be authorized (permitted) and all other cooling water systems, open or closed, must be registered in France (see also Section 8.1.2.3, Table 8.3 for summary).

The new French regulations require periodic environmental monitoring (surveillance) to determine the concentration of viable *Legionella* bacteria in the water of all cooling towers. Data from this activity is necessary to quantitatively document that control of the hazard has been achieved under operating conditions (validation). For an explanation of how validation fits into HACCP (Principle 7), see Section 6.3.6 and Chapter 8.

Periodic determination of viable *Legionella* in cooling water systems must be performed. The French standard plate count method for *Legionella* (AFNOR T90-431, see Section 8.1.2.1) is to be used by laboratories that have been approved (accredited) by the government. The environmental surveillance schedule to validate hazard control under operating conditions is specified in the regulation: monthly *Legionella* analyses for facilities subject to authorization (permits), bi-monthly for facilities subject to registration, and tri-monthly (quarterly) for facilities in which the results of the analysis are less than 1000 colony forming units per liter of cooling water (CFU/l) during 12 consecutive months. If the results of quantitative viable *Legionella* analysis in cooling water indicates that concentrations are greater than or equal to 1000 but less than 100,000 CFU/l, then the system must be immediately cleaned and disinfected; a new analysis to determine the viable *Legionella* concentration must be performed within at least 15 days; if concentrations remain in this range (1000–100,000 CFU/l), then a new hazard analysis and control plan with improvements must be established and repeated until the concentration is below 1000 CFU/l.

If the concentration of viable *Legionella* in the cooling water is shown to be greater than 100,000 CFU/l, then the facility owner/manager must immediately inform the local government authority and immediately shut down the facility for cleaning and disinfection according to the corrective action previously established. The owner/manager of the facility must place a notice for public view on the cooling tower which reads (paraphrased) "*Urgent and important – cooling tower exceeds limit of 100,000 CFU/l Legionella*". Before the facility can be restarted, the owner/manager must establish a new hazard analysis and control plan with improvements in both hazard control methods and verification (verification is proof that the plan is being implemented and is an important component of HACCP Principle 7, see Chapter 6). Environmental samples must be analyzed by the standard plate count method 48 h after restart of the facility and then at least every 15 days thereafter for

3 months. If during this period the concentration of *Legionella* is shown to be above 10,000 CFU/l, then the facility is to immediately shut down again, report to the local government authority and repeat the corrective action procedure before restarting the facility. In cases where shutdown of a facility poses grave risk, exceptions to the shutdown rules can be made by special permission from the government, but more frequent cleaning and disinfection and more frequent environmental surveillance of viable *Legionella* analysis using the standard method will be required.

If a case of legionellosis is associated with a facility or suspected to be associated with a facility, environmental samples must be immediately taken for analysis with the standard plate count method to be performed by an accredited approved laboratory. Isolated colonies of *Legionella* obtained from the cooling water system under investigation must be sent to the National Centre of Reference (CNR) in Lyon for genome identification of the strains of *Legionella*.

5.2.4.2 European industry-sponsored guidance

In Europe, Eurovent 9/5 and more recently 9/6 (Eurovent 2000) are *Legionella* control guidance documents that have been compiled by collaboration between the makers of cooling water equipment, treatment chemicals and treatment devices. The guidance is comprehensive and generally consistent with government-sponsored guidance adding the excellent benefit of pan-European perspective.

5.2.5 The United States of America

Although there are no US government standards or directives for preventing legionellosis, the Occupational Safety and Health Act of 1970 specifies the General Duty clause (Section 5, Duties) which can be applied to cover *Legionella* in poorly maintained water systems:

- Each employer shall furnish to each of his employees employment and a place of employment which are free from recognized hazards that are causing or likely to cause death or serious physical harm to his employees.

That there are penalties associated with violating the General Duty clause is given in Section 17, Penalties:

- Any employer who willfully or repeatedly violates the requirements of Section 5 of this act … may be assessed a civil penalty of not more than $70,000 for each violation, but not less than $5000 for each violation.

That legionellosis is a recognized hazard which can cause or is likely to cause death or serious harm to employees is given in the Occupational Safety and Health Administration (OSHA) Technical Manual and in also in the Code of Federal Regulations (OSHA 1998; CFR 2002).

The US Department of Labor published the widely-cited *OSHA Technical Manual on Legionnaires' Disease* in 1998. It gives guidance in regard to the legal responsibility of employers to protect the health and safety of employees, customers, and the surrounding community according to the *Occupational Health and Safety Act*. The *Code of Federal Regulations* (CFR Title 40 §142.52) in the USA gives the

Primary Drinking Water Standard and includes there should be zero detectable *Legionella* in drinking water.

The Centers for Disease Control and Prevention have published several guidance documents that refer to *Legionella* (CDC 2001, 2003, 2004). Three US states (Maryland, Pennsylvania, and Texas) have developed their own guidance for *Legionella* control and prevention of legionellosis (Pennsylvania 1997; Maryland 2000; Texas 1999). The most proactive of these is the guidance from Pennsylvania. As can be seen from the incidence rate data in Chapter 4 (Table 4.1), more legionellosis is detected per capita in Pennsylvania compared to any other and this is almost certainly due to proactive policy and better surveillance.

5.2.5.1 US industry-sponsored guidance

In the USA, industry-sponsored guidance is especially important in the absence of any one controlling guidance document from the Federal government. This industry-sponsored guidance taken together establishes a performance standard in the US to which facility owners/managers should comply in order to protect their employees, customers, and the community from legionellosis. Compliance to this guidance can mitigate an accusation of negligence. In a litigious society such as in the US, motive of this kind to comply with a code of practice can be extremely compelling.

Legionella control guidance in the US has been published by the American Association of Heating, Refrigeration, and Air-Conditioning Engineers (ASHRAE 2000), the Cooling Technology Institute (CTI 2000), the American Society of Testing and Materials (ASTM 2002), the Associated Water Technologists (AWT 2003) and others.

5.3 A GLOBAL CODE OF PRACTICE

Apparent in guidance documents from all over the world are general elements of good practice. Specific recommendations vary considerably but common elements

Table 5.2 Generally recommended practice to prevent legionellosis.

Recommended practice	Purpose
Risk management	Identify hazards, assess, or characterize risk to prioritize hazards that could harm people, identify improvement needs, assign management responsibilities for communication, training and for implementing a plan to prevent legionellosis
Control, control limits and monitoring control limits	Establish control measures, limits, monitoring, and specific corrective actions such as specifications for emergency disinfection
Environmental sampling for the pathogen	Document that the hazard has been reduced to acceptable levels, eliminated or prevented from harming people under actual operating conditions (validation)
Periodic audits and reassessment	Independently audit and periodically document that the plan is properly implemented (verification)

can be categorized into a general technical consensus opinion about how to prevent legionellosis (Table 5.2).

These practices for the prevention of legionellosis can be further categorized as follows:

- Hazard analysis (Chapter 6)
- Hazard control (Chapter 7)
- Validation and verification (Chapter 8)

6

Hazard analysis

Many hundreds of thousands of legionellosis cases could have been prevented by application of practical scientifically proven hazard analysis and control principles.

There are two *Legionella* hazard analysis and control systems recommended for use today: water safety plans (WSPs) and hazard analysis and critical control point (HACCP) plans. They are conceptually and functionally equivalent. Either system can be equally used to prevent legionellosis. One of these systems should be used in every facility.

Hazard analysis and control plans should be simple, cost-effective, and easy to implement. Otherwise, people will not use them.

6.1 PUTTING IT INTO PERSPECTIVE: WATER MANAGEMENT PLANS AND BUDGETS

Every facility should have a water management plan because the cost of water and energy has significantly increased in recent years and because there is a duty to provide hazard-free water. Buying water and then paying to dispose of it is a significant part of every facility budget. In many regions, the cost to dispose water to the sewer exceeds the cost to purchase the water. The energy cost of inefficiently operating water utilities is very high. For many facilities, the water system is assumed to be hazard free and therefore very little money, if any, is spent on water safety. Figure 6.1 conceptually illustrates components of the water management plan and budget.

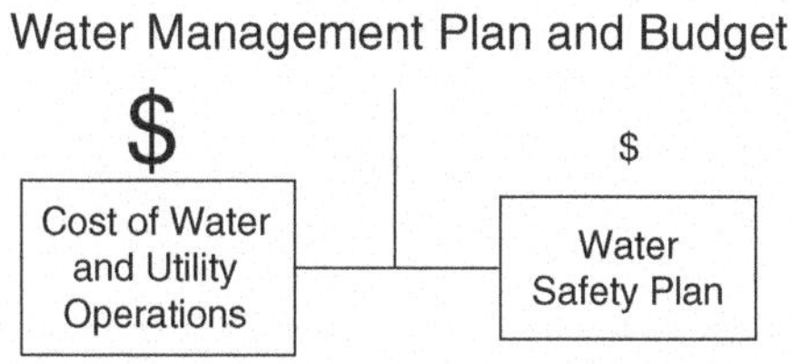

Figure 6.1 Every facility should have a water management plan and budget. All facilities spend money to cover the cost of water, its disposal, and the costs of running utilities. Most facilities spend little or nothing for water safety. This is the reason why hundreds of thousands of people are directly or indirectly affected by legionellosis each year.

Inefficiency costs in water-related operations are often highest when safety is poorest. This is especially true if water and energy waste is due to microbial fouling in the system. The best first step to improve water system safety is therefore to reduce inefficiencies in the system. The best next step is to allocate some of the savings to safety improvements. In Figure 6.1, the dollar sign above "water safety plan" can be increased at the expense of a decrease in the dollar sign above "Cost of water and utility operations". Allocation of the budget in this way is a practical means to economically justify reducing waste and increasing safety of facility water systems.

Practice to reduce inefficiencies in the water system and in the water treatment program are outside the scope of this book except to say that often it can be achieved through better use of water pre-conditioning, more economical and more effective water treatment practice, and closer attention to energy conservation.

Squarely within the scope of this book is how to efficiently and cost-effectively use the savings from better water management to improve the safety of the water system. This is achieved by hazard analysis and control.

6.2 THE WSP

The World Health Organization (WHO) has established a hazard analysis and control scheme for engineered water systems called the WSP (WHO 2004). The conceptual components of the system are: Establishment of a team to develop the plan, hazard analysis, risk characterization for each hazard, establishment of control limits, validation that control limits are effective, management systems for support, and independent surveillance to verify that the plan is implemented.

6.2.1 Using WSPs to prevent legionellosis

In its publication in preparation, *Legionella and the Prevention of Legionellosis*, the use of the WSPs hazard analysis and control system is highly promoted by the WHO (2004). The WSP is a scientifically based approach to hazard analysis and control. Figure 6.2 is a schematic representation of the WSP.

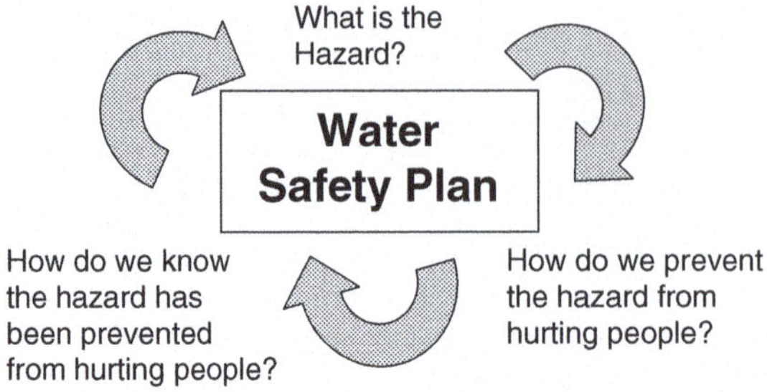

Figure 6.2 Schematic representation of the WHO hazard control scheme (WHO 2004).

6.2.2 Developing a WSP to prevent legionellosis

For hazard analysis and control schemes to be useful, they must be practical, easy to develop, and cost-effectively implemented. A systematic approach to developing the plan is given in Figure 6.3.

Prominently discussed in the WHO document on WSPs (WHO 2004) is the concept of validation which is the investigative work necessary to show that control limits are quantitatively effective under operating conditions in the system. This is usually achieved with well-directed use of pathogen analytical testing. Independent surveillance that the plan is being implemented (verification) is also prominently required in WSP (see also Chapter 8 Validation and Verification).

6.2.3 Risk characterization in the WSP system

After hazards have been identified, then they must be prioritized before control activities can be cost-effectively planned. This process is called risk characterization (*note*: not "risk assessment"). The severity and the likelihood (i.e. the probability) that a hazard could hurt people should be systematically determined. Tables 6.1 and 6.2 are examples of simple systems to prioritize the severity and probability of hazards (WHO 2004).

6.2.3.1 Assessing the risk of legionellosis

Risk assessment is required in many Codes of Practice as the first step in establishing a risk management plan (e.g. HSC 2000; DHS 2001a). In hazard analysis and control plans, risk assessment *per se* is not required. However risk characterization is necessary as shown in the previous section (see Figure 6.3). Although this distinction between risk characterization and risk assessment may seem at first merely semantic, upon closer examination it should become clear that it is not a trivial difference. Hazard analysis and control is a quantitative process with the goal to eliminate, reduce, or prevent the hazard (in this case, the hazard is *Legionella*) from harming people. Legionellosis risk assessment is a qualitative process which can at best help identify precautions that when implemented will *reduce the risk* of contracting legionellosis (for additional perspective about hazard analysis and control relates to risk, refer to Section 6.3.4). To see that legionellosis

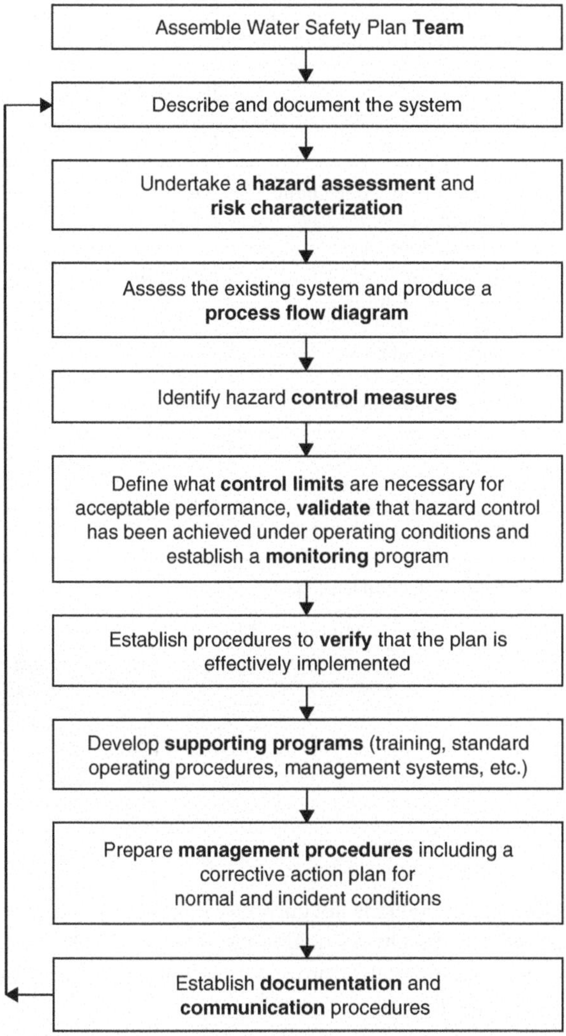

Figure 6.3 Steps in the development of a WSP as recommended by the WHO. Text in bold indicates key concepts in the WSP hazard analysis and control system.

risk assessment is a qualitative process, one must consider all that is required in order to perform a quantitative risk assessment.

To quantitatively assess risk, at least the following is required:

1. Dose–response data previously determined in well-controlled experiments for use in quantification of the potential of the hazard to cause harm.
2. Determination of the extent and duration of human exposure to the hazard.
3. Characterization of the possible consequences resulting from exposure (Percival *et al.* 2000).

Table 6.1 Criteria for hazard severity and risk (probability or likelihood) that the hazard could harm people for use in risk characterization scoring (adapted from WHO 2004).

Severity		Risk (probability)	
Categories	Criteria	Categories	Criteria
Insignificant	No impact or not detectable	Almost certain	Once per day
Minor	Potentially *harmful* to a small population	Likely	Once per week
Moderate	Potentially *harmful* to a large population	Moderate likely	Once per month
Major	Potentially *lethal* to a small population	Unlikely	Once per year
Catastrophic	Potentially *lethal* to a large population	Rare	Once per every 5 years

Table 6.2 Hazard severity/risk matrix for use in analyzing hazards and scoring risk (adapted from WHO 2004).

Risk (probability)	Severity of the consequence				
	Insignificant	Minor	Moderate	Major	Catastrophic
Almost certain					
Likely					
Mod. likely					
Unlikely					
Rare					

Since knowing these parameters is not currently possible for legionellosis, the risk assessment process must be always qualitative (Cooper 2004). Further, legionellosis risk assessment is a qualitative process as follows: permissible exposure levels are difficult to set for *Legionella* because:

1. the susceptibility of exposed persons greatly varies,
2. the mixture of biological contaminants is complex and varied,
3. interactions among all biological agents present in the environment render exposure assessments impossible.

However, a simplistic quantitative microbial risk assessment for this organism has been developed based on available data (Storey *et al.* 2004b). Both pure *Legionella* cultures and *Legionella* co-cultured with *Acanthamoeba castellanii* were considered. Inputs to the model included: thermal disinfection data, free and combined chlorine disinfection data, biofilm *Legionella* concentration, biofilm detachment (sloughing) effects and estimates of the quantity of water inhaled in aerosols during a 10 min shower. The infectious dose of *Legionella* was assumed to be 1 viable cell. The calculations suggested that 100 *Legionella* per square centimeter

Table 6.3 Legionellosis risk classification scheme (adapted from DHS 2001a).

Risk classification	Higher risk → Lower risk Critical risk criteria A	B	C	D
Critical risk factors	If the system matches any of the responses below.	Matches any of the responses below and none in A	Matches any of the responses below and none in A or B	Matches any of the responses below and none in A, B, or C
Stagnant water	System is idle more than a month. Recirculation pump not timed. Dead legs prevalent.	Same as A but with timed recirculation pump.	Any single factor in A	System operates continuously and there are no dead legs.
Nutrients and growth	Contaminated water. No corrosion control. Wet surfaces open to sunlight.	Any two factors in A	Any one factor in A	No factors in A
Poor water quality	No automated biocide dosing protocol. No water treatment program.	No automated biocide dosing. Water treatment program.	Automated biocide dosing. No water treatment program.	Both automated biocide dosing and water treatment program.
Poor equipment design or operation	No system design review. No system operation and performance review. No drift eliminators.	Drift eliminators for cooling towers in place.	Same as B with at least one review lacking.	Both reviews in place. Good drift eliminators.
Location and access	Located in a healthcare or residential care facility or large numbers of people are exposed.	Same as A but moderate numbers of people exposed.	Same as B but few people exposed.	Not located near susceptible people.

of biofilm in the absence of disinfection combined with a 10% sloughing event and inhalation of 50 μl of water each day corresponded to an annual risk of 1.0 (i.e. certain infection). In order to reduce this risk to the US Environmental Protection Agency (EPA) target risk level for waterborne disease of 1 infection per 10,000 people per year, a reduction in *Legionella* concentration of more than three orders of magnitude would be required.

It should be noted that although progress towards a quantitative risk assessment model has been made, the consensus opinion at this time is that legionellosis risk assessment must be considered qualitative.

6.2.3.2 Effectively using legionellosis risk assessments

It can be difficult to document progress in reducing legionellosis risk since the risk assessment itself is qualitative. Therefore, schemes have been designed to provide risk ranks and risk scores. Table 6.3 is adapted from the published code of practice in Victoria, Australia and provides an example of one way to rank legionellosis risk (DHS 2001a, b, 2002).

The higher purpose of risk classification (or characterization) and assessments should be to help organize specific hazard control recommendations. For example, if a system is ranked in the A category (higher risk), then more frequent service should be specified in pathogen analysis, cleaning with disinfection, and monitoring with control compared to a system that is assessed into a lower risk classification. Progress in reducing risk can be documented by "reassessing" the system after implementation of control measures. When improvements result in a lower risk classification for the system, then less frequent service to maintain control can be recommended.

Another approach to dealing with the qualitative nature of legionellosis risk is to use a scoring system (e.g. DHS 2001b). These schemes can be helpful especially for water systems in which population susceptibility to legionellosis is always high (e.g. the domestic water supply in hospital). The key to making risk scores work is to establish pre-determined criteria for each critical risk factor characterized and then, to apply these criteria uniformly across all systems. Risk classification activity should be part of hazard analysis and control plan (i.e. the WSP).

The WSP is a hazard analysis system that is conceptually equivalent to another system, the HACCP system, also endorsed by the WHO. This system is recommended globally as the preferred means to prevent foodborne hazards from harming people. It is endorsed by the WHO in its *Codex Alimentarius* (Food standards). The WSP and HACCP systems are functionally identical and equally effective. Every facility management team should use one or the other, WSP or HACCP, to prevent biological hazards from harming people.

6.3 THE HACCP SYSTEM

The HACCP system is a scientifically based method to prevent hazards from harming people. It has been proven effective for controlling biological hazards in the USA and globally. The system is conceptually simple, cost-effective, practical and quantitative.

In France, the HACCP system was specified in 2005 as the preferred hazard analysis and control system for preventing legionellosis (see Section 5.2.4.1 and France 2005b).

For prevention of waterborne disease, the HACCP system is being applied or considered for water systems all over the world including in Australia and New Zealand (Nadebaum *et al.* 2003), Belgium (Dewettinck *et al.* 2001), Finland (Horman *et al.* 2004), Germany (Kistemann *et al.* 2001), Italy (Angelillo *et al.* 2001), South Africa (Jagals and Jagals 2004), Sweden (Westrell *et al.* 2004), and the UK (Howard 2003).

At the time of this writing, the HACCP system is not yet widely used to prevent legionellosis disease in America. However, since so many kitchens in American facilities of interest (such as hospitals, hotels, and restaurants) are already preventing foodborne illnesses by implementing HACCP plans, it will be practical to extend those plans to prevent waterborne disease, including legionellosis, from their water systems.

Most HACCP plans at the food service level in America are developed voluntarily. Many restaurants and food service kitchens (such as in hospitals) routinely implement HACCP plans to prevent foodborne pathogens from causing disease in their facilities. Development of HACCP plans in the food processing/manufacturing industry is driven either by a legal mandate, or by buyers of foods who require suppliers to have such systems in place. HACCP systems are means of assuring safe foods and ingredients in the processing sector.

Many HACCP plans to control foodborne pathogens in America are required by regulation published in the Code of Federal Regulations (9 CFR 417). The United States Department of Agriculture (USDA) requires food production plants to use HACCP. The Food and Drug Administration (FDA) requires seafood processors to use HACCP. The FDA also requires HACCP plans in certain canned-goods production processes to prevent botulism.

In response to quantitatively documented success in the USA, the WHO has adopted HACCP as the preferred recommended system for preventing foodborne disease worldwide.

6.3.1 HACCP system principles

The system is conceptually simple. It comprises 7 principles:

1. Conduct a *hazard analysis.*
2. Identify *critical control points* (CCPs) (the last control point to prevent, eliminate, or reduce the hazard).
3. Establish *critical limits* for each CCP.
4. Establish a *monitoring* plan for critical limits at CCPs.
5. Establish *corrective actions* for each critical limit.
6. Establish procedures to *document* all activities and results.
7. Establish procedures to confirm that the plan actually works under operating conditions (*validation*), is being implemented properly (*verification*) and is periodically *reassessed.*

In practice, a few preliminary steps are necessary before these seven principles can be applied. They are as follows:

- Assemble an HACCP team including one HACCP-trained person (e.g. a certified HACCP Manager/Consultant, a trained employee or other resource, e.g. from a local college).
- Identify the use and users of the product (in this case, water), especially to determine at-risk consumers.
- Identify all uses for the product (in this case, water) at the facility.
- Develop process flow diagrams to describe how the product is processed in the facility.
- Verify that process flow diagrams are accurate by on-site audit.

It is advised that there should be at least one HACCP-trained person on the HACCP team who is independent of the facility owner or management.

An excellent authoritative source for detail is the *Guidebook for the Preparation of HACCP Plans* available online at www.fsis.usda.gov/Science/HACCP_Models/index.asp

6.3.2 The HACCP plan and supporting documents

HACCP plan development should be easy and not excessively time consuming. Implementation of the HACCP plan must be practical and cost-effective. *In lieu* of this, people will not voluntarily use the system.

Typically, there are five kinds of HACCP-related documents for a facility water system (Table 6.4). They are the process flow diagrams, the hazard analysis summaries, the validation criteria, the verification/validation schedule, and the HACCP plan document.

Example process flow diagrams for a hypothetical water system are given in Figures 6.4 and 6.5. Example hazard analysis summaries for the potable water system and for the utility water system are given in Tables 6.5 and 6.6, respectively; these hazard summaries correlate to the process flow diagrams in Figures 6.4 and 6.5. Validation criteria for the hypothetical potable water system are given in Table 6.7. A validation and verification schedule plan for the hypothetical water system is given in Table 6.8 (validation and verification are further discussed with more detail in Chapter 8). An example HACCP plan document for this hypothetical water system is given in Table 6.9. The figures and tables in this chapter can be used as templates for any water system but since each system is different, modifications will always be necessary.

6.3.3 Process flow diagrams

The first task for the HACCP team is to describe the products, how they are processed and how they are used. The water in every facility is processed for use in the utilities (heating and cooling) and in the domestic water service (cooking, drinking, bathing, showering, decorative fountains, etc.). For example, utility water used for heating and cooling may be processed by pre-conditioning with equipment such as softeners or filters, chemically treated to maximize operational efficiency,

Table 6.4 Supporting documents and the HACCP plan for a facility water system.

Documents	Description	Purpose
Process flow diagrams for each product (the potable water and utility water)	Simple schematics to show how water is processed in the facility. Water in every facility is processed for utilities and for domestic (potable) service. Examples of water-processing steps are conditioning, softening, filtering, storage, cooling, heating, distribution, cleaning, disinfecting, etc.	Identifies the products (utility water or potable water) and the processing steps used to prepare water for use in the utilities (e.g. the boilers, cooling towers, fire-suppression systems) and for use or consumption by building occupants (e.g. cooking, drinking, bathing, showering, etc.).
Hazard analysis summaries	Systematic list of hazards for each processing step. Indicates the severity/risk of each hazard. Lists the specific control measures that could be applied to prevent, eliminate, or reduce the hazard.	Identifies CCPs for the process. Typically, there will be one or two CCP for each process.
Validation criteria	Lists the specific quantitative measurement criteria for each hazard or for performance criteria directly related to the hazard.	Evidence (data) to confirm that the hazard has been eliminated, reduced to acceptable levels or prevented under operating conditions from harming people.
Validation and verification schedule	Systematic list of required activities, frequencies, responsibilities, the review schedule and the dates. Assignment of responsibilities and accountability is critical to successful implementation of HACCP plans.	Coordinates activity to obtain data proving that the plan actually works (validation) and that the plan is being properly implemented (verification). Coordinates documentation of control limit monitoring and corrective action reports.
The HACCP plan document	Lists the specific control limit for each CCP, the monitoring frequency, the corrective action required if a critical limit has been breached, reporting responsibilities and verification schedule.	Provides details of the entire plan in one document. During an HACCP audit, this is the key document for review.

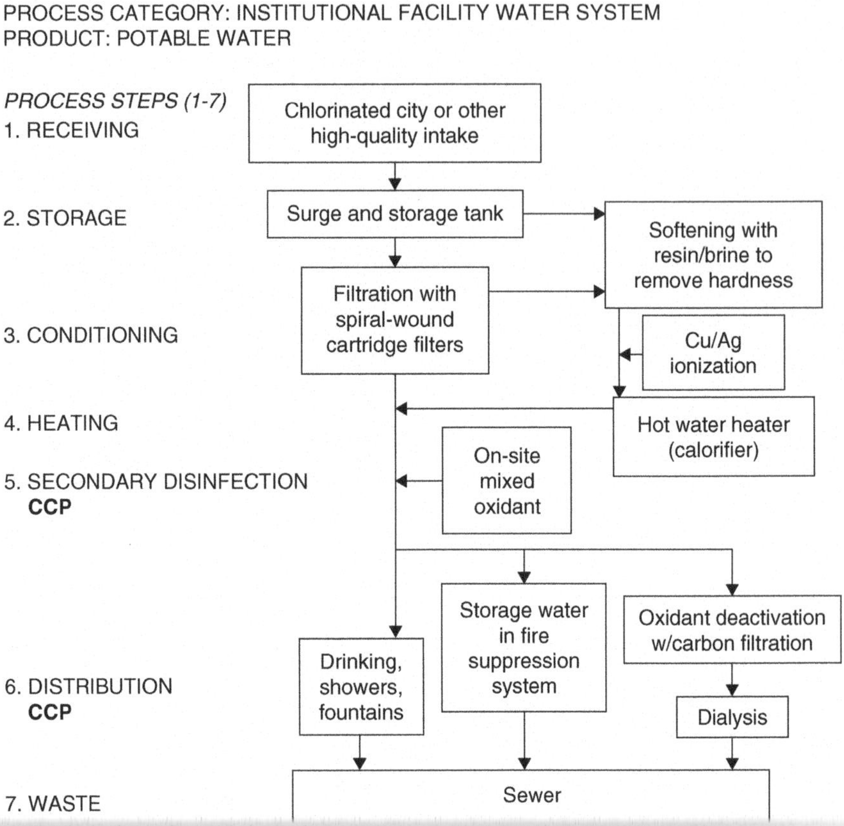

Figure 6.4 Process flow diagram for a hypothetical domestic water service system in a hospital with CCPs indicated.

stored, distributed, and recirculated. Potable water for the domestic water service may be processed by softening, filtering, storing, heating, disinfecting, and distributing, for example. Each facility processes water uniquely but there are of course basic similarities in every water system.

The process flow diagram is a simple schematic that conceptually shows step-by-step in numbered sequence how the water is processed in the facility. It is not a detailed or complex plumbing schematic. After the team drafts the process flow diagram it must be verified by an on-site inspection. The completed and verified process flow diagram is required before hazard analysis and risk assessment can begin. An example process flow diagram for an institutional facility water system is given in Figures 6.4 and 6.5. Note that CCPs are indicated on the process flow diagram. They would have been identified in the hazard analysis step.

6.3.4 Hazard analysis and risk assessment

Hazards are biological, chemical, physical or radiological entities that have the potential to cause harm. The objective of an HACCP plan is to prevent, eliminate

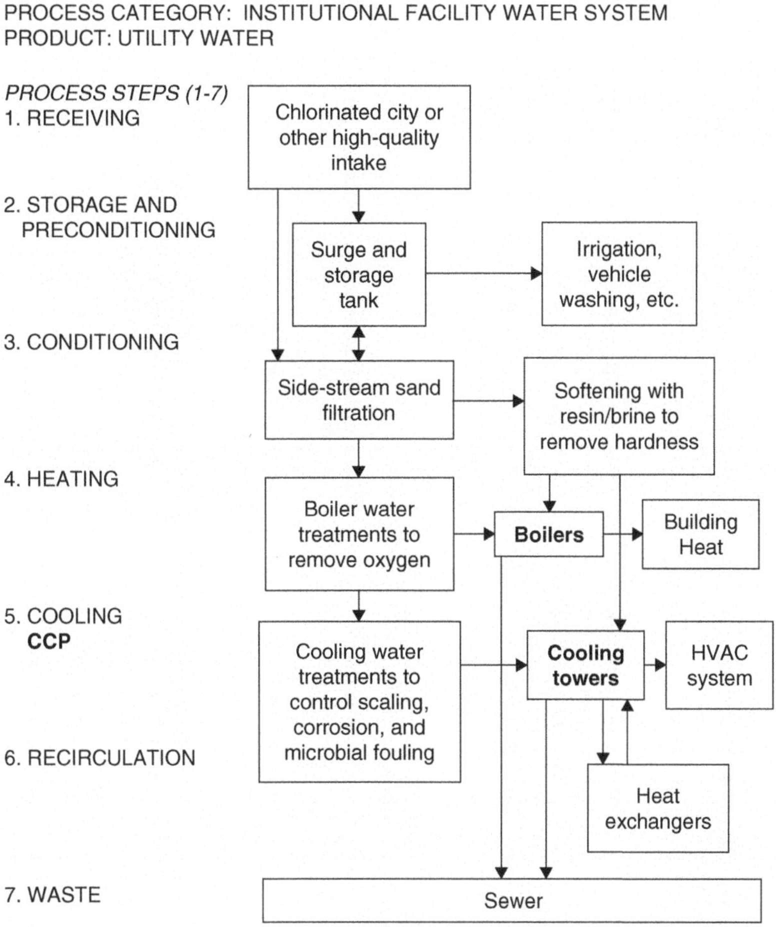

Figure 6.5 Process flow diagram for a hypothetical utility water service system in a hospital with CCP indicated.

or reduce hazards. For example, a drinking water distribution system in a hospital is, by definition, not a hazard. However, if the drinking water system is not properly maintained, then it could become contaminated with *Legionella*, which is a known biological hazard. These statements are also true for cooling towers and other components of facility water systems.

Risk is the likelihood (or probability) that a hazard could hurt people. Although quantitative data can be used to assess the risk that a hazard could hurt people, experts say legionellosis risk assessment is a qualitative process (Cooper 2004). Recently, attempts have been made to quantify the risk assessment process, but extremely simplistic models are required (Storey *et al.* 2004b, refer to Section 6.2.3.1).

Plans driven by risk assessment connect control activities to qualitative assessments of risk. The goal of risk assessment-driven control is to reduce the risk that hazards will hurt people. A claim to eliminate risk is not appropriate because risk

Table 6.5 Example hazard analysis summary for the hypothetical potable water system shown in the process flow diagram (Figure 6.4).

Product: potable water	Identify potential hazard introduced, enhanced, or controlled at this step	Risk severity: risk significant?	Basis for the risk characterization	What controls *could* be applied to eliminate, reduce, or prevent the hazard from causing harm?	CP	CCP
Processing steps						
1. Receiving	B: Biological hazards such as coliforms, *Legionella*, virus, and protozoa C: Chemical hazards such as heavy metals or disinfection by-products P: Radioactive contamination hazards	No	Moderate to low risk because purchased water is treated to US standards for drinking water given in the CFR (2002)	1. Obtain product from sources that are certified to the National Drinking Water Regulations (NDWR) 2. Obtain water quality test results from the water provider every 6 months	1B 1C 1P	
2. Storage	Currently no on-site storage	No	N/A	N/A		
3. Conditioning	B: Biological hazards from growth in filter media and on ion exchange media	No	Moderate to low risk because filter and softeners are regularly serviced and well maintained	1. Backwash filtration media regularly 2. Disinfect filtration media 3. Monitor microbiological fouling of filtration media 4. Regularly service ion exchange system	2B	
4. Heating	B: Biological hazard from growth of microorganisms in the hot water heater P: Physical hazard from scalding	Yes	Water heaters have been associated with severe disease from pathogens such as *Legionella*. Other potentially pathogenic microorganisms can also grow in hot water heaters	1. Maintain 60°C (140°F) in the hot water heater continuously 2. Drain and flush twice a year to remove scale and sediment 3. Treat continuously with Cu–Ag ionization (see Step 5, Secondary disinfection)	3B 2P	

5. Secondary disinfection	B: Biological hazards from pathogenic microorganisms are controlled in this step C: Chemical hazard from overdosing disinfectant	Yes	Must strictly maintain disinfectants at approved levels; the biological hazard could result from too little disinfectant and the chemical hazard could result from too much	1. Apply Cu/Ag (0.2–0.4 and 0.02–0.04 mg/l, respectively) for secondary disinfection at the hot water heater continuously 2. Apply mixed oxidant continuously (0.4–0.6 mg/l free residual oxidant) from electrolytic halide oxidation for secondary disinfection in the distribution system		1B 1C
6. Distribution	Biological hazards will accumulate on plumbing surfaces, pipe walls, showerheads, and in dead legs	Yes	Microbial pathogens have been found to grow on surfaces in biofilms and can re-enter the water	1. Deliver 50°C water at distal taps within 1 min at full on hot 2. Regularly flush all taps weekly 3. Remove all dead legs		
7. Waste	Biohazards from fecal coliforms and viruses	No	Sewage can transmit waterborne pathogens	1. Ensure physical separation of potable water from sewage	4B	

CP: control points.

Table 6.6 Example hazard analysis summary for the hypothetical utility water system shown in the process flow diagram (Figure 6.5).

Product: utility water	Identify potential hazard introduced, enhanced or controlled at this step	Risk severity: risk significant?	Basis for the risk characterization	What controls *could* be applied to eliminate, reduce, or prevent the hazard from causing harm?	CP	CCP
Processing steps						
1. Receiving	B: Biological hazards such as coliforms, *Legionella*, virus, and protozoa C: Chemical hazards such as heavy metals or disinfection by-products P: Radioactive contamination hazards	No	Moderate to low risk because purchased water is treated to US standards for drinking water given in the CFR (2002)	1. Obtain product from sources that are certified to the NDWR 2. Obtain water quality test results from the water provider every 6 months	1B 1C 1P	
2. Storage and pre-conditioning	B: Biological hazards such as coliforms, *Legionella*, virus, and protozoa	No	The water storage/surge tank in this facility is used intermittently	1. Remove the water storage tank 2. Apply microbial control to the storage/surge tank	2B	
3. Conditioning	B: Biological hazards from growth in filter media and on ion exchange media	No	Moderate to low risk because filter and softeners are regularly serviced and well maintained	1. Backwash filtration media regularly 2. Disinfect filtration media 3. Monitor microbiological fouling of filtration media 4. Regularly service ion exchange system	3B	
4. Heating (boilers)	P: Physical hazard from scalding	Yes	Steam under pressure	1. Proper boiler maintenance and pre-treatment is required	2P	

5. Cooling (cooling towers)	B: Biological hazards from pathogenic microorganisms are controlled in this step	Yes	Pathogenic microorganisms such as *Legionella* can proliferate in poorly maintained	1. Control makeup water hardness 2. Control pH 3. Use scale inhibitors 4. Use corrosion inhibitors		1B
	C: Chemical hazard from overdosing biocides and other inhibitors		cooling water systems Transmission of the pathogen to susceptible people can cause severe life-threatening disease	5. Use industrial biocides to control microbial fouling 6. Perform regular maintenance to keep system clean 7. Maintain drift eliminators in good condition		1C
6. Recirculation	B: Biological hazards will accumulate on plumbing surfaces, pipe walls, and cooling tower surfaces	Yes	Microbial pathogens have been found to grow on surfaces in biofilms and can re-enter the water	Same as for Step 5	4B	
7. Waste	Possible biohazards from fecal coliforms and viruses Possible chemical hazard from overfeeding inhibitors	No	Sewage can transmit waterborne pathogens	1. Ensure physical separation of utility water from sewage 2. Ensure no overfeed of inhibitors and biocides	5B 2C	

CP: control points.

Table 6.7 Example validation criteria correlated to the hypothetical water system described in Figures 6.4, 6.5 and in Tables 6.5–6.9.

Processing step	CCP (number)	Validation criteria	Frequency
Product: potable water			
5. Secondary disinfection	1B	1. Heterotrophic plate count <500 CFU/ml at hot water heater	Quarterly
	1C	2. Less than 10 CFU/ml (not detectable) viable *Legionella* in the standard method at hot water heater	Quarterly
6. Distribution	2B	3. Heterotrophic plate count <500 CFU/ml at 10 distal sites	Quarterly
		4. Less than 10 CFU/ml (not detectable) viable *Legionella* by the standard method at 10 distal sites	Quarterly
Product: utility water			
5. Cooling (cooling towers)	1B 1C	*Corrosion control validation*: 1–2 ml/year on mild steel coupons in-line 0.1–0.2 ml/year on copper coupons in-line	Quarterly
		Scale control validation: No visible scale in heat exchangers	Quarterly
		General microbial control validation: total count by dip slides < 10E5 CFU/ml	Weekly
		Legionella *control validation*: Less than 10 CFU/ml (not detectable) viable *Legionella* by the standard method at hot water heater	Quarterly

Note: All validation criteria and frequencies listed in this table are for a hypothetical water system. They should not be construed as general recommendations for any particular water system nor can they supersede local or regional guidance and regulation.

can never be entirely eliminated. Nor is it advisable to claim that risk has been minimized to an extent that makes it insignificant. Risk management professionals will advise that the goal regarding risk should be to substantially *reduce* it.

Site-specific risk assessments are a requirement in some legionellosis control guidance documents for building water systems. The UK Approved Code of Practice and Guidance (L8) for the control of *Legionella* is the most striking example (HSC 2000). Similarly, risk assessment-driven *Legionella* control guidance is given by governments in many other countries (see Chapter 5). Risk assessments produced in these countries are often subject to supplier-based limitations which may be conflicted in regard to services and products offered for control of *Legionella*.

The term "risk assessment" does not appear anywhere in the official guide for developing HACCP plans nor in CFR for HACCP System (9 CFR 417, US Code of the Federal Register). The first HACCP principle (conduct a hazard analysis) requires that the severity and the risk of each identified hazard must be determined. This is where risk assessment (or more properly called "risk characterization") fits into HACCP.

Table 6.8 Example validation and verification schedule correlated to the hypothetical water system described in Figures 6.4, 6.5 and in Tables 6.5–6.9.

Activity	Frequency	Responsibility	Reviewer
Initial validation of the HACCP Plan	Prior to and during initial implementation	Independent surveillance	General Manager/Owner HACCP consultant
Verification that CCP monitoring is according to plan	Once per shift	Engineering technician	Executive Engineer
Subsequent validation of the HACCP plan	Change in critical limit Change in equipment Change in process System failure One time per quarter	Independent surveillance HACCP consultant	General Manager/Owner
Review of corrective action monitoring to verify that it is according to plan	Two times per year	Executive Engineer	General Manager/Owner HACCP consultant
Scheduling verification activities	One time per year or if system changes	Executive Engineer	General Manager/Owner
Comprehensive HACCP plan verification and reassessment	One time per year	Executive Engineer HACCP consultant	General Manager/Owner

In this context, risk characterization adds information useful for identifying hazards and determining their significance. Risk characterization helps the HACCP team to understand factors influencing the likelihood that the water system may be contaminated; it does not however lead directly to the most effective control measures. The key to ensuring that risk assessment is not used to motivate ineffective or useless service is to make certain that CCPs are properly identified. This is accomplished through a method of prioritizing risk as part of the hazard analysis (for applicable methods to prioritize risk, see Section 6.2.3). Operators do this before establishing CCPs. Operators therefore ensure the efficient allocation of resources in the HACCP plan. The hazard analysis must be very detailed to be successful in establishing CCPs. Competent individuals must identify all risks logically and categorize risks as to their likely occurrence. The establishment of CCPs is based on sound science, not guess work or assumptions. Hazards and their priorities must be evaluated for each water system based on a site survey. There can be no one-size-fits-all hazard analysis.

Quantitative pathogen analytical tests for *Legionella* can be effectively used to identify and analyze hazards in a water system. However, negative results from these tests do not prove there is no hazard due to the limitation of the sampling and analytical methods. Therefore, although they can be helpful, pathogen analytical tests are not required for the hazard analysis.

The purpose of the process flow diagram and the hazard analysis is to identify CCPs.

6.3.5 CCPs

Hazard analysis-driven plans connect hazard control activities to the last opportunity in the process at which it is quantitatively possible to prevent, eliminate, or reduce hazards from harming people. These control opportunities are called CCPs.

Hazard control is achieved by setting specific control limits for CCPs and monitoring them. If a deviation from the control limit occurs, then corrective actions are immediately implemented. Figure 6.6 is a decision tree useful for determining if control at a process step is critical and should be therefore identified as a CCP.

Note that CCPs are indicated in the process flow diagrams but that is just for convenience and reference. The basis for deciding CCPs comes from the hazard analysis. During hazard analysis, the HACCP team identifies control points for each hazard. Hazards are classified as biological, chemical, or physical (including radiological hazards). Not all control points are critical. The CCPs are to be indicated on the hazard analysis summary sheet. This systematic analysis of hazards is to be well documented. Tables 6.5 and 6.6 give an example of an HACCP hazard analysis summary based on the hypothetical water system for which process flow diagrams are given in Figures 6.4 and 6.5.

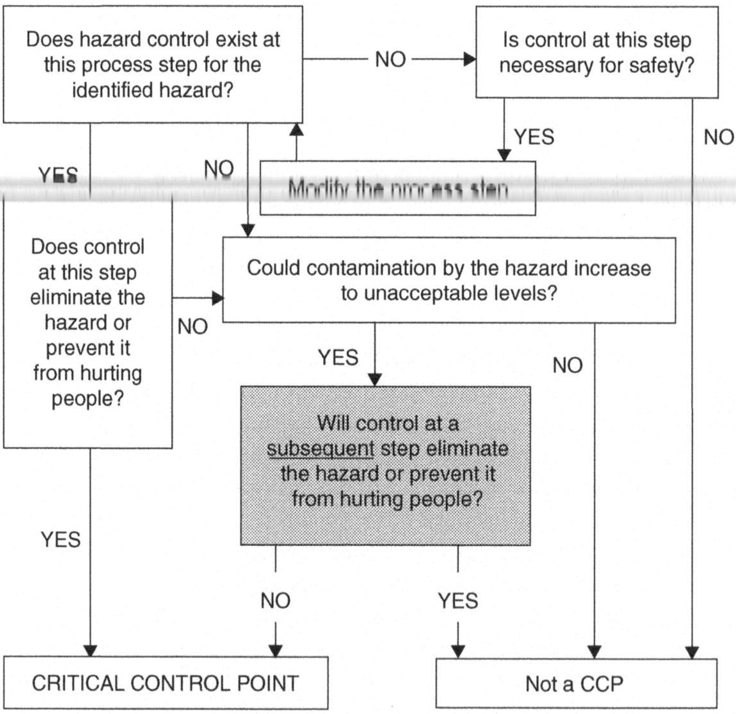

Figure 6.6 Decision tree for determining if a process step is a CCP. The shaded box places emphasis on the fact that a control point is not critical if control at subsequent steps in the process can eliminate, prevent or reduce a hazard to acceptable levels.

6.3.6 Validation, verification, and reassessment of the plan

As with the WSP, quantitative confirmation that control limits are effective against the hazard (validation) and independent audits to confirm that the plan is being properly implemented (verification) are essential requirements in HACCP.

For an example that correlates with the hypothetical system under discussion, control limits could be validated in this example as shown in Table 6.7. Note that no performance criteria are indicated for any of the CCPs. This is because performance under operating conditions is validation of hazard control and should not be confused with control limit monitoring (see Chapter 8 for more detail).

After deciding validation criteria to prove that the hazard has been controlled under operating conditions, the HACCP team should establish the validation and verification schedule. Table 6.8 is an example which correlates to the hypothetical water system in discussion.

6.3.7 The HACCP plan document: control limits, monitoring, frequency, corrective actions, and verification schedule

The only control points that require limits and monitoring are the CCPs. This is where the power and cost-effectiveness of the HACCP system is made manifest. Critical control limits, monitoring, frequencies, corrective actions, and verification responsibilities are established for each CCP and listed in on the HACCP plan. This is the controlling document for the entire plan. As an example, an HACCP plan for the hypothetical system in Figures 6.4 and 6.5 and Tables 6.5–6.8 is given in Table 6.9.

Further given as a hypothetical example for discussion, if the hot water heater had been identified as a CCP, the water temperature control limit specification might be set at 60–65°C at all times. An example of a control limit for disinfection of a domestic water service CCP might be not less than 0.5 and not more than 0.8 mg/l (ppm, w/v) free residual oxidant at selected sentinel outlets all times. In the utility water system, an example of control limits for a cooling tower CCP might be a chemical concentration range measured for scale, corrosion, and biological inhibitors.

Note that control limits are not based on measurements of the hazard (such as the concentration of *Legionella* in the water). This is because the purpose of control limits is to set parameters for applying the control method. Control limits and monitoring do not directly prove nor disprove that the hazard has been eliminated, prevented, or reduced to acceptable levels under operating conditions (validation).

In HACCP plans, control limits are routinely monitored and the results are carefully documented. For example, a logbook of water heater temperature measurements and disinfectant concentrations may be all that is required for monitoring a system with two CCPs. Control limits are usually quantitative, but they can also be semi-quantitative or semi-qualitative.

In HACCP, corrective actions are specified for every CCP in the event that there is a deviation from the control limit (refer Table 6.9). For example, a cooling tower may be identified as a CCP with control limits set for chemical concentrations of microbial inhibitors (i.e. the biocides). If the control limits are out of specification, then a corrective action to physically clean and disinfect the cooling system might be required as one of the corrective actions. As another example, if a domestic water service heater is a CCP and it is allowed to cool below the control limit, then thermal disinfection of the potable water system might be given as one of the corrective actions. It is important that HACCP documentation includes a log of any corrective actions taken. This corrective actions log should correspond to documented deviations in CCP control limits. That is, for every documented devation in a CCP control limit, there must be a documented corrective action or documented explanation given.

The general progression for developing an HACCP plan is to:

1. establish the team;
2. produce process flow diagrams (e.g. Figures 6.4 and 6.5);
3. perform hazard analysis to identify the CCPs (e.g. Tables 6.5 and 6.6);
4. establish the validation criteria that will quantitatively prove that the hazard has been controlled under operating conditions (e.g. Table 6.7);
5. establish the validation and verification schedule (e.g. Table 6.8 and see also Chapter 8);
6. for each CCP list critical control limits, monitoring, frequencies, corrective actions and verification schedule/reporting in the HACCP plan document (e.g. Table 6.9).

Quantitative proof (validation) that the CCP limits can eliminate, prevent, or reduce the *Legionella* hazard to an acceptable level under operating conditions is necessary in HACCP. Although methods of validation are still evolving, there must be proof that the control limits set at CCPs are scientifically valid. HACCP experts are increasingly recommending microbiological testing under operating conditions to assure that critical limit values are in fact eliminating, preventing, or significantly reducing microbiological hazards. Validation, a key element at the seventh principle of HACCP, should not be confused with routine monitoring of control limits at CCPs. For an example of validation, refer to the new French regulations which require environmental sampling and pathogen analysis for validation that the *Legionella* hazard has been controlled in cooling towers (see Section 5.2.4.1).

6.3.8 Disease prevention with HACCP

The HACCP system has been quantitatively proven to prevent disease in the USA. Statistically significant decreases in disease have been documented by the Centers for Disease Control and Prevention (CDC) since 1996, the year in which the final HACCP rule was published. From 1996–2003, *Salmonella typhimurium* infections, the leading cause of foodborne disease in the USA, decreased by 38%. Infections of *Yersina* declined 49%, *Escherichia coli* O157 by 42%, *Campylobacter* by 28%. From 1997 to 2003, the incidence of *Cryptosporidium* infection decreased 51%. (Preliminary FoodNet data on the incidence of infection with pathogens transmitted commonly through food – Selected Sites, USA, 2003. *Morb Mort Weekly Report* **53**(16): 338–343).

Based on quantitative evidence that the system is effective in the real world, HACCP is now being applied to preventing legionellosis and other waterborne diseases for the first time in the USA.

In 2005, the HACCP system was specified as the preferred hazard analysis and control method for preventing legionellosis in France (2005b).

A study of HACCP history is helpful for perspective about the system and how it can be applied to preventing legionellosis.

6.3.9 A history of the HACCP system

Development of the HACCP system began in 1959 by the Pillsbury Company in response to the US National Aeronautical and Space Administration (NASA) request for assurance of foodborne illnesses prevention. By 1971, HACCP was the preferred method for controlling foodborne illness in the space program. Since then it has been gradually accepted as an effective process to prevent disease. In 1973, the FDA incorporated HACCP into regulations to control botulism. Studies sponsored by the USDA evaluated more effective and less expensive ways to improve food inspections. These studies resulted in recommendations to shift responsibilities from food inspectors in the Food Safety Inspection Service (FSIS) to facility management such as is required in the HACCP system. This recommendation was negatively received by many food inspectors (a unionized work force), food producers, and consumer groups. Very little progress was made in the 1980s but the National Academy of Sciences recommended use of the system in 1985.

By 1992, the USDA was aggressively recommending improvements due to a tragic outbreak of illness in the Northwestern USA caused by *E. coli* O157 : H7 in under-cooked ground beef. In 1996, the finalized HACCP standard was published by the FSIS, a branch of the USDA. The first regulatory reference for HACCP appeared in 1997 (Code of the Federal Register, November 28, 1997, Vol. 62, No. 229). In 1998, a scandalous outbreak of listeriosis occurred. Management at the offending meat processing plant eventually pled guilty to covering-up prior knowledge of *Listeria monocytogenes* contamination in food shipped to several states. The death toll had reached 14 when the USDA publicly warned that 55 million Americans were at risk. This resulted in political action. Since 1998, USDA requires all food processing plants in the USA to use HACCP (9 CFR Part 417). The requirement applied in establishments with 500 or more employees on January 26, 1998, in establishments with 10 or more but fewer than 500 employees (unless annual sales are less than \$2.5 million) and on January 25, 2000 in establishments with fewer than 10 employees or annual sales of less than \$2.5 million.

These regulations apply to food processing plants only. Most retail establishments apply HACCP voluntarily. They do this to create a due diligence defence, reduce liability, and to improve operational efficiencies or improve regulatory compliance. Major restaurant chains, hospital kitchens, and hospitality (hotel) kitchens are examples of voluntary users of the HACCP system. They use HACCP plans to protect employees, customers, and the surrounding community from pathogen hazards. In the kitchens of most restaurants in America there will be a food safety-trained person on duty during every shift not just because of regulations but rather, because oversight of the food safety system is required.

In the major restaurant and hotel franchises that voluntarily use HACCP, line cooks on every shift and other kitchen personnel are responsible for specific control

Table 6.9 Example HACCP plan for the hypothetical water system described by process flow diagrams (Figures 6.4, 6.5) and hazard summaries (Tables 6.5, 6.6).

Processing step	CCP number	Critical control limit	Monitoring	Frequency	Limit deviation corrective actions	Records	Verification procedure
Product: potable water							
5. Secondary disinfection	1B 1C	1. Cu/Ag concentration: 0.2–0.4 and 0.02–0.04 mg/l, respectively	Field test for Cu Laboratory test for Cu/Ag	Daily at hot water heater Quarterly	Check flow cells, power source and electrodes; call supplier for technical support	Daily in secondary disinfection logbook	Engineering Technician, daily Executive Engineer, weekly
		2. Free residual mixed oxidant concentration: 0.4–0.6 mg/l continuously	Field test for free residual oxidant (FRO)	Daily at mixed oxidant generator	Check feed water, brine tank, flow cells and power source; call supplier for technical support	Daily in secondary disinfection logbook	Independent verification 2 times per year
6. Distribution	2B	FRO: 0.2–0.4 mg/l continuously	Field test for FRO	Daily at three distal sites previously selected	Check mixed oxidant generator; recheck distal sites; notify management	Daily in secondary disinfection logbook	Same as above
Product: utility water							
5. Cooling (Cooling Towers)	1B 1C	1. Water hardness and 50–75 mg/l as $CaCo_3$ equivalently 2. Water pH: 7.5–8.0	Field test alkalinity alkalinity Field test water pH	Weekly hardness and Weekly	Check operation of softeners, filters, chemical feed pumps, product levels, mixed oxidant on-site generator	Daily in water treatment logbook	Engineering Technician, daily Executive Engineer, weekly

3. Inhibitors: 8–10 mg/l phosphate; 2–4 mg/l zinc; 3–5 mg/l azole	Field test inhibitors	Weekly	Check feed water, brine tank, flow cells and power source; call supplier for technical support	Daily in water treatment logbook	Independent verification 2 times per year
4. Biocides: 0.5–1.0 mg/l tritium oxide (TRO) mixed oxidant generated on-site	Field coupons	Quarterly			
15–20 mg/l non-oxidizing biocide shots M, W, F	Field test for FRO	Daily			

Note: All critical limits, monitoring and frequencies listed in this table are hypothetical and should not be construed as general recommendations for any particular water system. Each water system is different. Water treatment programs should be established and maintained by specialists.

activities as specified in the HACCP plan. There is a growing trend for all employees in the establishment to understand HACCP principles and to be aware of how pathogen hazards are controlled. In some major restaurant chains, such as the Darden Corporation, even the food servers know about HACCP and what the acronym means. At more than 1300 locations in the USA and Canada, Florida-based Darden Corporation is the leading operator of casual-dining restaurants, including flagships Red Lobster and Olive Garden. Red Lobster is the No. 1 seafood chain (about 650 units) and Olive Garden leads the Italian-dining segment (about 540 outlets). Verification that the plan is being correctly implemented is documented by an independent reviewer/inspector at regular intervals (usually 2 times per year) in every restaurant. This is evidence to the significance of voluntary HACCP risk management implementation in the USA.

Hospitals and other medical facilities have been leaders in the implementation of HACCP. Dieticians usually receive training in HACCP in their education making them familiar with the concept. Pressure from accrediting bodies such as the Joint Commission on Accreditation of Hospitals (JCAHO) to use risk management systems for infection control are the reasons we see HACCP systems in almost every hospital dietary department in the USA.

The Joint Commission for the Accreditation of Healthcare Organizations (JCAHO) issued a new standard that became effective from January 1, 2001. This standard, EC 1.7, requires all JCAHO accredited facilities to have a management program to "reduce the potential for organizational-acquired illness". It holds healthcare facility management responsible for "managing pathogenic biological agents in cooling towers, domestic hot water, and other aerosolizing water systems".

Furthermore, in the JCAHO Environment of Care – Equipment and Infection Control (PI 320) sections, proactive risk assessments are specified for compliance. HACCP systems comply with this requirement, but this is delegated to the contracted dietary management, or other environmental management company in most cases.

At the food processing level, verification of the HACCP plan has taken on new dimensions as HACCP systems mature. Not only are auditors looking for accurate record keeping and validation of critical controls, auditors are also assessing management commitment. ISO standards are now being integrated into HACCP verification activities. This trend toward ISO is buyer driven. Buyers want to know that suppliers are not just filling out forms correctly and filing them. Buyers are insisting that management systems assure the sustainability of the HACCP system. Documented, accredited HACCP training and certification of key management and quality assurance (QA) staff is also emerging at the verification principle.

6.3.10 A cost–benefit analysis of HACCP in USA

Since 1998, HACCP plans have been implemented in at least 6200 food production facilities in the USA. The cost–benefit analysis has been favorable.

The ongoing total cost of hazard management to control listeriosis and other foodborne pathogens is about \$357 million which is about \$0.01 per pound of food produced in the USA. These costs include creating pre-requisite programs such as general sanitation and food safety training and HACCP-related costs such as analyzing hazards, identifying CCPs, establishing control limits, monitoring control limits, corrective actions, verifying control, and documentation of all activities. Most of the cost of HACCP is related to upgrading equipment, structural repairs, training, and labor costs associated with more extensive cleaning, better maintenance, and more pathogen analytical testing. For example, before 1998, there were about 3500 *Listeria* tests run per year in the USA. This rate has steadily increased to 10,000 tests run per year in 2003. Reduced costs (i.e. savings) from medical expenses and lost time due to implementation of risk management plans in the USA food industry have been estimated at $1 to 3.7 billion. (McCluskey, J. 2004. Opportunities to market food safety. Washington State University, http://classes.arec.wsu.edu/550/, from data obtained in Caswell 1994.)

6.3.11 Biological hazard control in the food industry

Before HACCP, the government bore responsibility for policing plant sanitation practices and implementing actions through the FSIS. Today, the HACCP system puts food safety in the hands of plant management to be implemented by their own employees with help from their independent service providers. The role of government inspectors in HACCP is to review the plans and verify that they are implemented correctly. This is a fundamental change in responsibilities.

Not everyone likes the HACCP system. The Union of federal employees working for the USDA complained that before HACCP, responsibility for testing, regulating, specifying and in some case implementing sanitation in food plants was theirs. Now their role is to check that HACCP plans are in place and to verify that control has been achieved. This has resulted in a loss of responsibility for these people.

The most serious criticism of HACCP is that some plant managers are cutting corners by using a "one plan fits all" mentality. There is evidence of cases in which this is proven to have occurred. Such abuses of the system result in failure to do the right things at the expense of doing HACCP paperwork just to fulfill the "letter of the law". This problem certainly exists where government regulation is the only driving factor, but where the buyer is the voluntary driving factor, facilities risk losing their ability to sell their products if they fail to implement validated systems or falsify records. Therefore it is quite rare for HACCP systems audited by independent third parties to be just rote paperwork exercises.

7

Hazard control

Important note: Hazard control should be applied only after first developing a systematic hazard analysis and control plan for the facility (see Chapter 6). Failure to implement control measures within the context of a risk management plan could result in greater exposure to the hazard or cause additional hazards to occur. The first rule in hazard control is "do not make matters worse". The best way to avoid breaking the first rule in hazard control is to develop a plan before any control methods are used. Even if emergency disinfection procedures are deemed necessary, a basic plan should be developed before implementation.

Disclaimer: This chapter is not inclusive of all available hazard control technologies nor is it intended to recommend any one technology over another. Neither is the chapter inclusive of all systems in which hazard control may be necessary. The reader is referred to government-sponsored or approved guidance and industry-sponsored guidance in his/her region (refer to Chapter 5 and Table 5.1) for further information and local regulation.

The biological hazard caused by *Legionella* can be eliminated, prevented or reduced to an acceptable level in facility water systems. Applications of physical, chemical, and biological control methods are usually necessary.

For most facilities, there are two separate water processes: Utility water (i.e., heating and cooling water) and potable water (i.e., domestic water service). By far, most legionellosis is associated with exposure to contaminated potable water systems whilst many of the most highly publicized outbreaks of disease have been associated with utility water systems such as cooling towers.

The recommendations given in this Chapter are derived from published or otherwise communicated opinions of technical experts in various regions and countries. These recommendations do not supersede local or regional regulations, standards or Codes of Practice. The reader is referred to published guidance in his/her locality (see Table 5.1) to make certain that control measures taken are in compliance with government-approved practice.

Generally, pathogen proliferation can occur in nearly all water systems. Many factors are involved but most importantly are the presence of microbial biofilms, the degree of microbial diversity, and the availability of nutrients especially in warm water systems. Managing the microbial fouling process to prevent legionellosis principally consists of controlling biofilms and limiting microbial diversity within the entire system. Exposure to *Legionella* occurs by aspiration (deep inhalation into the lungs) of microscopic (<8 μm diameter) aerosols, contaminated water or infective particles. Certain industrial water systems produce aerosols as an inherent function of their operation. Minimizing aerosol dissemination from such systems can directly reduce health-related risk. Limiting exposure of susceptible people to the hazard focuses on the proximity of aerosol-generating systems to human populations. Although the most susceptible individuals to legionellosis are elderly males, heavy smokers, alcoholics, patients with chronic pulmonary disease, and immunosuppressed people (Arnow *et al.* 1982; Bollin *et al.* 1985; Addiss *et al.* 1989; Breiman 1993; Bentham *et al.* 2002; McCoy 2003b), the demographics and extent of the disease now make it quite clear that there is a very wide range of at-risk people who do not fall into those categories (see Chapter 4).

7.1 HAZARD CONTROL IN POTABLE WATER SYSTEMS

Most legionellosis occurs from exposure of susceptible people to contaminated potable water systems. All potable water systems in buildings are comprised of two systems that work together: The central system (heating devices and distribution pipe work) and the peripheral systems (outlets such as showers, faucets, and taps).

Building water systems can become contaminated with *Legionella* if the water is allowed to stagnate especially at the optimum growth temperature range of 30–40°C. Risk management plans for these systems should place emphasis on maintaining good water circulation and avoidance of optimum growth temperatures.

7.1.1 Hazard control in domestic hot water systems

"Domestic" refers to all non-utility water used for lavatories, showers, drinking fountains, etc., in commercial, residential, hospitality, healthcare, and industrial settings. These systems are often linked to legionellosis disease. Transmission of the pathogen from domestic hot water is by aspiration (inhalation) of *Legionella*-contaminated water or aerosolized microscope water droplets or infective particles.

Water heaters in healthcare, hospitality and industrial settings frequently contain cool zones near the base where cold water enters and scale and sediment accumulate. Dead legs or periodically stagnant pipe runs (such as infrequently used showers and taps) often provide areas in which *Legionella* can proliferate to dangerous levels.

Recommendations for control (after first establishing a hazard analysis and critical control plan, see Chapter 6):

- Store hot water at a minimum 60°C (140°F) and deliver at a minimum of 50°C (122°F) to all outlets (see Figure 7.2). *Caution*: Beware of scalding hazard; fail-safe scald-protection equipment such as pressure-independent, thermostatic mixing valves at delivery points should be used to regulate delivery temperatures.
- The hot-water tank should be drained periodically (twice a year at least) to remove scale and sediment and cleaned with chlorine solution (5–10 mg/l total residual oxidant), if possible. The tank must be thoroughly rinsed to remove excess chlorine before reuse.
- Point-of-use water heaters can eliminate stagnation of hot water for infrequently used lines and should be used wherever possible.
- Hot-water lines should be insulated and hot water temperatures in the safe range should be maintained in the distribution lines.
- Eliminate dead legs wherever possible. Locate and eliminate areas in the hot water system below 50°C (122°F). Frequent flushing of hot water lines will also reduce the hazard.
- Drain and flush showers and taps regularly. Install self-purging shower systems wherever possible especially for highly susceptible people (see Section 7.1.4.2 for details about regular flushing and self-purging shower systems).
- Domestic hot-water recirculation pumps should run continuously. They should be excluded from energy conservation measures.
- Many domestic hot water systems require disinfection to control waterborne pathogen hazards including *Legionella* (see Section 7.1.3 for details). The most effective disinfectants for these systems are copper–silver ionization, chlorine, chlorine-based mixed oxidants generated on-site electrolytically, chlorine dioxide and chloramines.

7.1.1.1 Emergency disinfection of domestic hot water systems

Emergency disinfection is costly and potentially hazardous. It should be therefore avoided. The best way to avoid emergency disinfection is to establish and implement an effective water safety plan (Chapter 6).

In the case of an outbreak or detectable pathogen in a domestic hot water system or detectable *Legionella* from swabs taken at taps or showerheads, thermally disinfect by increasing the water-heater temperature to 70°C (158°F) for 24 h or to the maximum temperature allowable by regulation. Remove aerators on the faucets and then flush each outlet for at least 20 min. Clean and disinfect the aerators in 10 mg/l chlorine bleach solution before reinstalling. Flush *all* taps with hot water because contaminated stagnant areas will reinfect the system. *Caution*: Beware of the potential for serious scalding hazards.

Chemically disinfect hot water systems by hyperchlorination at the tank to produce at least 10 mg/l free residual chlorine from bleach or from on-site electrolytically generated oxidant and flush *all* taps until a distinct odor of chlorine is evident. Maintain at least 2 mg/l free residual chlorine throughout the system for at least 2 h but not to exceed 24 h. It may be necessary to add as much as 20–50 mg/l free chlorine at the water heater to achieve this concentration of disinfectant at distal sites. Control of pH between 6.0 and 8.0 is recommended to ensure disinfection efficacy. Thoroughly flush the system after disinfection. *Caution*: Chlorine is corrosive and will shorten the service life of metal plumbing. Avoid the need for emergency disinfection.

7.1.2 Hazard control in domestic cold water systems

Domestic cold water systems can be a source of *Legionella* especially if cold-water lines are not maintained consistently below 20ºC. Even in very cold systems such as ice machines in hospitals, elevated levels of *Legionella* have been observed because the pathogen is not killed by low temperatures.

Dental water lines are common sources of microorganism-laden water including *Legionella.* Operating conditions are especially conducive to *Legionella* proliferation because the water is stagnant much of the time, the narrow plastic tubing encourages biofilm formation, and the water temperature is maintained at or near 37°C (98.6°F) for the comfort of the patient. Aspiration of dental rinse water is a potential source of *Legionella* infection. Recommendations: Filter water at the point of use with replaceable, inline, 0.22-microns pore sizes filters to prevent this hazard from harming patients and staff in a dental facility.

Water tanks that allow water to remain uncirculated for long periods can also promote growth of bacteria. They should be eliminated or designed to reduce storage time to one day or less. They should be covered to prevent contamination and protected from temperature extremes. In the UK for example, cleaning and disinfection of these cold-water storage tanks is required (HSC 2000).

Cross-contaminations of the domestic cold-water system with other systems should always be suspected. All connections to process water should be protected by a code-approved plumbing device (e.g., back-flow preventer, air gap, etc.). If significant contamination of the domestic cold water system occurs, the source of contamination must be determined. Inspect the system for dead legs and areas in which water may stagnate. Elimination of these sections or frequent flushing of taps to drain stagnant areas may be necessary. Insulate cold-water lines that are close to hot-water lines to prevent heat transfer.

7.1.2.1 Emergency disinfection of domestic cold water systems

Emergency disinfection is costly and potentially hazardous. It should be therefore avoided. The best way to avoid emergency disinfection is to establish and implement an effective water safety plan (Chapter 6).

Emergency hyperchlorination can eliminate, reduce or prevent *Legionella* in cold water systems from hurting people. Free chlorine levels of at least 2 mg/l from chlorine bleach or from on-site electrolytically generated oxidant should be maintained for at least 2 h but not to exceed 24 h throughout the system. It may be necessary to add as much as 20–50 mg/l free chlorine at the cold water storage tank. Run all faucets and taps until the odor of chlorine is present. Control of pH between 6.0 and 8.0 is recommended to ensure disinfection efficacy. Thoroughly flush the system after disinfection. *Caution*: Chlorine is corrosive to pipes. Emergency disinfection of cold-water systems should be avoided except in systems where high levels of *Legionella* have been observed or if an outbreak has occurred. Be aware that local regulations or Codes of Practice supersede the recommendations given in this chapter.

7.1.3 Hazard control in heating ventilation and air-conditioning systems

Heating, ventilation and air-conditioning (HVAC) systems are usually dry and are not therefore often sites in which *Legionella* can proliferate because the organism cannot survive without water. However, HVAC systems can disseminate contaminated water aerosols. Certain parts of the system (such as condenser drain-pans and humidifiers) are wet. The following control measures should be considered for HVAC systems:

- Eliminate or minimize use of water reservoirs, sumps, and pans. Chemically untreated, stagnant, warm-water sources provide an ideal environment for *Legionella* growth.
- Drain water sumps when not in use; for example, install an electric solenoid valve on the sump drain.
- Provide a "bleed" for water sumps so that dissolved solids do not accumulate as sediments in the sump.
- Slope sumps from the bottom so that all water can drain out and so that the pan can dry.
- Drain all inactive sumps to prevent accumulation of sediments and biohazardous substances. Maintenance failures and ambient warming (e.g., by sun light) can produce contaminated, stagnant water.
- Locate HVAC fresh-air intakes so that they do not draw aerosols from cooling towers, evaporative condensers, or fluid coolers.
- Use steam or atomizing humidifiers instead of units that use recirculated water. Do not use raw steam from the central heating boiler because it contains corrosion inhibitors and anti-scaling chemicals. Contaminant-free water must be used in atomizing humidifiers.
- Test all HVAC equipment periodically to insure that it is performing efficiently and as designed.

7.1.4 Hazard control in the peripheral water system

The peripheral water system in buildings consists of showers, taps and faucets some of which can be stagnant for long periods of time creating dead leg conditions. Conventional showers and faucets with mixer valves are thought to be the single largest source of exposure to *Legionella* bacteria. Showers and faucets not in regular use present a reasonably foreseeable risk to health. Government-approved Codes of Practice and other guidance describe two control procedures (e.g., HSC 2000): regular flushing with monitoring of conventional showers and installation of self-purging shower systems.

7.1.4.1 Conventional showers: regular flushing and monitoring

The hazard associated with proliferation of *Legionella* in peripheral parts of the domestic water system such as dead legs may be reduced or eliminated by regular (at least weekly) flushing. When outlets are not in regular use, weekly flushing for at least 5 min can significantly reduce the number of *Legionella* discharged from the outlet. This control procedure has to be sustained as lapses can result proliferation of *Legionella* at the outlet. Hazard analysis and risk characterization may indicate the need for more frequent flushing where there is a more susceptible population: for example in hospitals, nursing homes etc.

When such an outlet is flushed, a series of events then follows:

- Contaminated stagnant water from the peripheral system (shower and associated dead legs) is displaced by water from the central system; this is the cooler unwanted water found at the beginning of a shower.
- Flushed water carries away waterborne *Legionella* and shears unstable biofilm lining the internal aspects of pipe work and fittings. Contaminated water and biofilm now passes from the shower but passage may be interrupted by internal devices such as the showerhead spray plate. Conventional spray plates are often associated with the highest levels of contamination in domestic water systems.

Recommendations for regular flushing of shower heads are as follows:

- Flush all shower heads at maximum hot water setting for at least 5 min at least weekly. Check to see how long before hot water arrives at the outlet; it should take no longer than 1 min to reach 50°C at the tap.
- Flushing of flexible tap tails and shower hoses should be done with the shower head placed on the shower floor. Hydraulic shocks and physical movement of the hose can dislodge biofilms that may not be completely expelled if the showerhead is clipped up. The next user may be exposed unless the water is fully drained from the flexible hose.

Caution: Flushing showers and faucets is potentially hazardous and has been associated with disease in people performing the operation. Personnel engaged in these control practices should be trained to avoid exposure to aerosols. Personnel should be not in highly susceptible risk categories such as smokers, diabetics, chronic lung or kidney disease, immunosuppressed (e.g., caused by steroid therapy) or alcoholics.

7.1.4.2 Self-purging shower systems

Manual weekly flushing of showers may be difficult to implement and expensive. An alternative solution should be considered: the self-puring shower system. This system is a means of immediately purging to drain contaminated water from within the shower and associated dead legs before use and without the production of aerosols. This approach is featured as a practical alternative to flushing and monitoring in *Legionella* control videos produced by the UK's Health and Safety Executive.

The Safepurge™ self-purging shower system (Figure 7.1) provides automatic peripheral system control for domestic water systems.

7.1.4.3 Whirlpool spas and hot tubs

These facilities have been associated with many cases of legionellosis and should be avoided by highly susceptible people. Whirlpool spas and hot tubs are subject to local and regional public swimming and bathing regulations. However, these devices are not swimming pools and must be operated, treated and maintained differently.

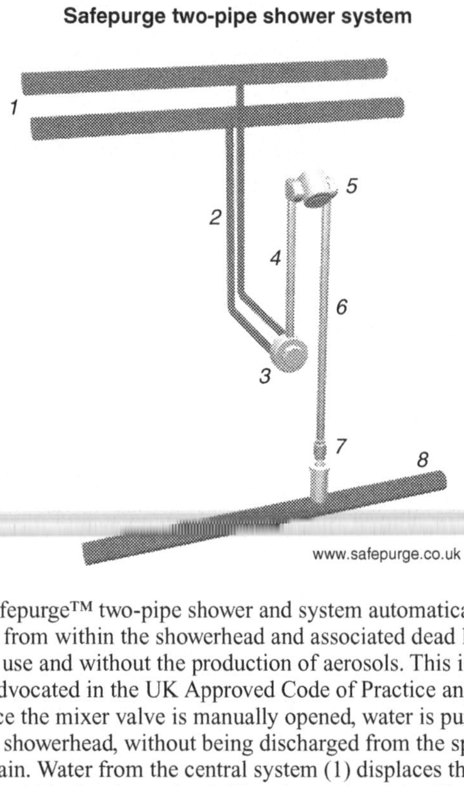

Figure 7.1 The Safepurge™ two-pipe shower and system automatically purges contaminated water from within the showerhead and associated dead legs to drain, immediately before use and without the production of aerosols. This is a *Legionella* control procedure advocated in the UK Approved Code of Practice and Guidance (HSC 2000, L8: 166). Once the mixer valve is manually opened, water is purged automatically; straight through the showerhead, without being discharged from the spray plate, and via the purge pipe to drain. Water from the central system (1) displaces the contaminated stagnant water from within the showerhead (5) and associated dead legs (2, 4); when the water temperature rises sufficiently to indicate an adequate purge has taken place, the purge cycle controller located at the end of the purge pipe (7) closes. Water is now redirected within the showerhead and a traditional spray pattern is delivered. The shower is to be operated in the conventional manner, adjusting temperature water to meet personal comfort requirements. The main visual difference between this shower system and conventional systems is the additional purge pipe (6). The person taking the shower will operate the mixer valve (3) in the normal manner by adjustment to a comfortable temperature setting; there will be a few seconds delay whilst the unwanted cooler stagnant water, found at the beginning of a conventional shower, is purged safely to drain (8) via the purge pipe (6). This system avoids manually flushing conventional showers weekly (as is recommended in several codes of practice), eliminates the need to monitor and document flushing procedures and also it avoids hazards from subsequent lapses in flushing. Artwork provided by Tim Jones (www.safepurge.co.uk).

The following recommendations are not intended to supersede local regulations to which the reader is also referred for operation of whirlpool spas and hot tubs.

- Restrict access by posting warnings that people who are immunocompromised or have chronic lung disease should not use the facility.
- Control bather load by posting and enforcing the maximum number of occupants. Typically, a circular 2.5 m diameter spa would have a maximum bather load of five persons at any given time.
- Filtration should be carefully controlled to the manufacturers' specifications. *Caution*: Failure to properly maintain spa filters has resulted in very serious legionellosis cases (see Section 1.3).
- Water chemistry should be carefully maintained. Typically, 4–6 mg/l free chlorine or bromine and pH 7.5 should be maintained continuously.
- Microbiological levels should be measured according to frequencies given in local or regional regulations. There should be less than detectable (by the Standard Methods for viable microorganisms) fecal coliforms, *Pseudomonas aeruginosa*, or *Legionella* and the heterotrophic place count (HPC) should be less than 200 CFU/ml.
- Regular daily maintenance of these devices should include hyperchlorination (10 mg/l free residual chlorine from bleach or electrolytically generated oxidant) for at least 1 h. The spa water should be replaced at least once a week. Daily water changes may be necessary if the spa is heavily used. When the water is changed, the spa should be physically cleaned including scrubbing of exposed surfaces, weirs, skimming devices, etc. to remove biofilm.
- Training of personnel in the proper operation and maintenance of these devices should be mandatory. Good recordkeeping of all operations is critically important.

7.1.4.4 Decorative fountains

These architectural features have been associated with serious cases of legionellosis and should be avoided by highly susceptible people. They should be eliminated from healthcare facilities, if possible. Decorative fountains should be operated and maintained to potable water standards, at least, for the locality. Recommendations given here are not intended not to supersede local or regional regulations:

- Water filtration or, for small systems, changing out the water should be done regularly at least once in every 2 weeks at least. The water should be clear (turbidity <1.0 NTU).
- Microbial fouling should be controlled with government-approved biocides. Chlorine residuals from bleach or electrolytically generated oxidant are very effective in these systems at 2–4 mg/l. Non-oxidizing biocides registered for decorative fountains are effective but safety issues should be carefully managed.
- Microbiological levels should be periodically monitored for decorative fountains in close proximity to susceptible people. There should be less than detectable (by the standard methods for viable microorganisms) fecal coliforms and *Legionella*; the HPC should be less than 500 CFU/ml.

Temperature °C	Temperature °F	Temperature effects	Thermal disinfection and inhibition recommendations
70	158	Disinfection occurs in minutes	**Emergency disinfection:** *Increase hot water heater temperature setting to >70°C and flush all taps for 20 min at 70°C or greater*
50	122	Disinfection occurs in hours	**Hot water storage:** *60°C (140°F) or higher*
40	104	Optimal growth	**Hot water system maintenance:** *50°C at point of use and hot water return* *Caution: Avoid scalding hazard*
30	86		
20	68	Slow/no growth	**Cold water storage:** *20°C or less inhibits proliferation but does not kill* Legionella

Figure 7.2 Thermal effects on *Legionella* bacteria are the basis for useful disinfection and growth inhibition recommendations.

7.1.5 Thermal disinfection and inhibition of growth

Legionella bacteria are sensitive to temperature effects (Figure 7.2). Above 50°C, rapid killing occurs that can be an effective means of disinfection. Below 20°C growth is inhibited to an extent that can prevent proliferation of the pathogen.

In practice, thermal methods are helpful but not reliable long-term for hazard control. Practical problems limiting thermal control of the hazard are maintaining target temperatures at distal sites, difficulty in maintaining practical flushing schedules and risk of scalding when water temperatures exceed 50°C. The most effective use of thermal disinfection is probably as a critical control point in hot water heater or storage systems (refer to Figure 6.1).

7.1.6 Secondary disinfection of potable water systems

Water delivered through the distribution system to the consumer will typically have been disinfected at the water treatment plant (primary disinfection). Responsibility for water quality transfers to the new owner of the water when the purchased water crosses the meter into the facility. The facility owner/manager must ensure that water quality is maintained through to the end user of the product. It is often necessary to further disinfect domestic water systems. This is called secondary disinfection or supplemental disinfection.

The water purchased will have been previously treated to potable water standards (primary disinfection) before entering the distribution system. In the USA, primary drinking water regulations include goals of zero (below detectable) *Legionella* (EPA 1999; CFR 2002), less than 500 CFU/ml heterotrophic aerobic bacteria and no detectable colifoms, viruses or *Giardia* (a protozoan parasite of mammals). In buildings with complex water distribution systems and hot water systems (which are, of course, nearly all buildings), microbiological quality can significantly degrade if disinfectant residual is depleted and if biofilms develop in the system and/or in dead legs.

Control in building water systems is best achieved by maintaining residuals of EPA-approved secondary disinfectants: chlorine, chlorine dioxide, chloramines or with EPA-approved disinfection devices or combinations of these preferred treatments throughout the facility with alternative disinfectants including copper–silver ionization, ozone and UV irradiation. Selection criteria and application guidance for secondary disinfection have been given (EPA 1999).

The most effective disinfection treatments to control *Legionella* in potable water systems are copper–silver ionization (Section 7.1.3.1), chlorine and electrolytically generated mixed oxidants (Section 7.1.3.2), chlorine dioxide (Section 7.1.3.3), chloramines (Section 7.1.3.4), and supplemental disinfectants including ozone (Section 7.1.3.5) and UV irradiation (Section 7.1.3.6) and other disinfectants (7.1.3.7).

Treatments have been compared in controlled laboratory experiments to simulate domestic water service real-world conditions (Thomas *et al.* 2004; Loret *et al.* 2005) and performance has been validated from surveys of field use (e.g., Lin *et al.* 1998; Neuner 2002; Craven 2003; Hall *et al.* 2003; Heffelfinger *et al.* 2003; Srinivasan *et al.* 2003; Stout and Yu 2003b; Sidari *et al.* 2004). All of these disinfectants have been shown to be effective when used properly.

Control and monitoring plans for the disinfectant applications are necessary for effective hazard control (Chapter 6) to ensure that the hazard is eliminated, prevented or reduced to an acceptable level under operating conditions.

7.1.6.1 Copper–silver ionization

In healthcare facilities, copper–silver ionization has the longest history and the most validation evidence (quantitative data showing performance under operating conditions) of *Legionella* hazard control. Control of *Legionella* in thousands of hot water systems has been achieved with copper–silver ionization worldwide dating back to the late 1980s. In a survey of 16 hospitals that had reported nosocomial legionellosis cases before installation of copper–silver ionization in hot water systems, there were no further reported cases of legionellosis after up to 11 years and positive *Legionella* indications were greatly reduced (Stout and Yu 2003b). Copper and silver ions are effective antimicrobials (EPA 2001) and they appear to act synergistically with chlorine (Lin *et al.* 2002).

Typically, copper–silver alloy electrodes (Figure 7.3) are ionized into the hot water system of the potable water to 0.2–0.4 and 0.02–0.04 mg/l of copper and silver cations, respectively. These concentrations are far below allowable levels for copper and silver in drinking water in the United States. The treatment is especially effective in hot water systems because temperature effects do not diminish antimicrobial performance of the metal ions.

To maintain precise copper–silver ion residual, the system must incorporate closed loop supervisory control whereby the computerized electronic control unit automatically adjusts output current based on water consumption or copper levels measured by a flow meter or copper analyzer. Systems can be configured, controlled, and monitored from either the local control panel or remote locations such

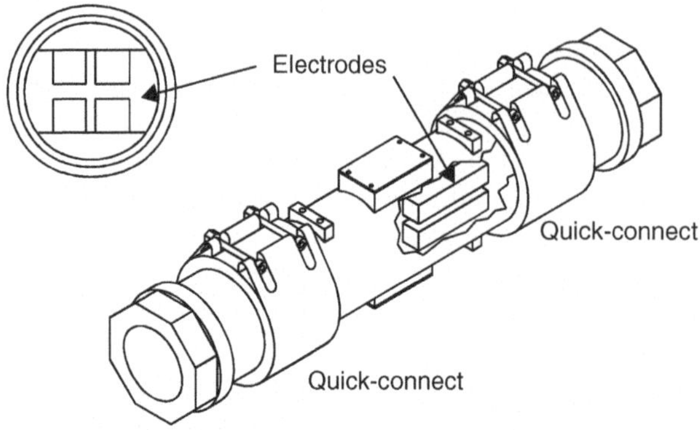

Figure 7.3 Copper–silver ionization has been effectively used to control *Legionella* and other waterborne pathogens in thousands of potable water systems all over the world. The ionizers are usually installed inline to the hot water heater (calorifier). This model (QLTF58-/4 Flow Cell, LiquiTech, Inc., Bollingbrook, IL) is 28.25 inches between the female-threaded adapters at each end. The ionizer itself is 16.25 inches long (between the quick-connects). The copper–silver alloy electrodes are shown in cut-away view. Also integral to the effective use of copper–silver ionizers is the controller (not shown). Artwork provided by LiquiTech, Inc. (www.LiquiTech.net).

as a building management system. Additionally, a direct modem interface can also provide remote supervisory and control functions.

Precautions regarding antimicrobial performance of copper–silver ionization systems are that effectiveness is reduced above pH 8 and the electrodes must be maintained free of scale or other deposits (Lin *et al.* 2002; Stout and Yu 2003b).

7.1.6.2 Chlorine and mixed oxidants

Microbial control using chlorine is the most often used and oldest method for disinfecting water systems. At use concentrations, chlorine is effective due to chlorination reactions with the organic molecules in the cell membranes of microorganisms; these reactions lead to rapid loss of viability. Feed and control systems for liquid chlorine (bleach solutions) and gas chlorine are very well developed (EPA 1999) and therefore not further discussed here.

An alternative to liquid or gas chlorine is on-site electrolytic generation of chlorine from sodium chloride (common salt). A popular and field-proven device to provide on-site electrolytic halide oxidation is given in Figure 7.4. Although oxidation of chloride to chlorine is the principle reaction, other oxidants are produced in this process at very low concentrations. In recent years, this on-site approach has become increasingly popular because of improved safety (no handling of concentrated chlorine) and better performance (mixed oxidants are more effective than chlorine alone, Figure 7.5).

The superior antimicrobial performance of on-site electrolytic generation of mixed oxidant solutions have been demonstrated in academically-oriented studies

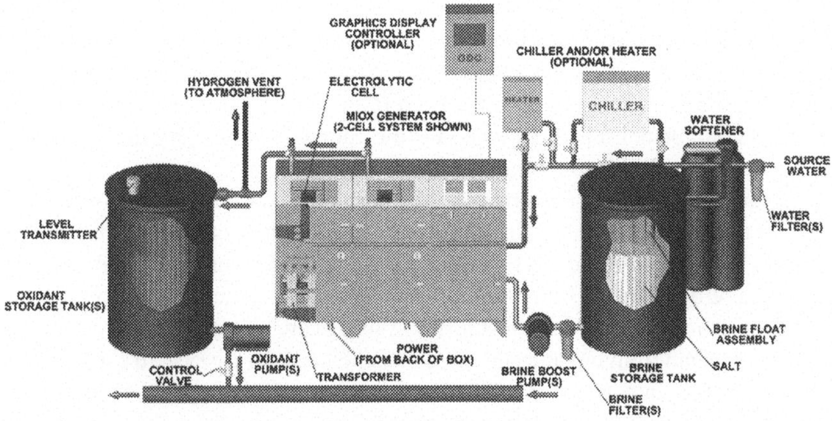

Figure 7.4 On-site electrolytic oxidation is an effective means to control waterborne pathogens in potable and utility water. The chlorine-based mixed-oxidant solution is produced from sodium chloride brine (common salt) in the electrolytic generator housed in the cabinets (center of the graphic). Calcium and other cations that impart hardness and will impede the process of electrolytic halide oxidation are removed from the feedwater to the generator with ion exchange resin (right). The feedwater is cooled or heated as necessary before entering the electrolytic generator from the brine makeup tank to the right. Mixed-oxidants produced in the electrolytic generator are transferred to the mixed-oxidant storage tank to the left of the generator where the mixed-oxidants are then fed into the water supply to disinfect the water. Artwork courtesy of MIOX Corporation (www.miox.com).

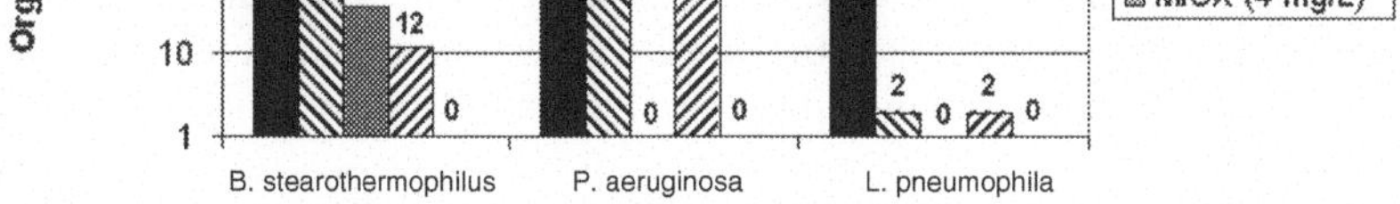

Figure 7.5. Antimicrobial performance of mixed oxidants from electrolytic oxidation of salt compared to sodium hypochlorite after 10 min of exposure in well-controlled laboratory experiments. Mixed oxidants caused orders of magnitude more inactivation of *Bacillus stearothermophilus*, *Pseudomonas aeruginosa*, and *L. pneumophila* than did equivalent concentrations of sodium hypochlorite. Mixed oxidants caused undetectable levels of *P. aeruginosa* and *L. pneumophila* at the 2 mg/L dose and of all microorganisms organisms at the 4 mg/l dose. Data and artwork courtesy of MIOX Corporation (www.miox.com).

and in practical real-world field applications (Shetty *et al.* 1999; Zinkevich *et al.* 2000; Thantasha 2002; Walker *et al.* 2003). Mixed-oxidant antimicrobial solutions generated on-site are sometimes referred to as "electrochemically activated (ECA) water" or "superoxidized water".

Treatment parameters for chlorine and mixed oxidants from electrolytically generated on-site halide oxidation are typically set at 0.5–1.0 mg/l (ppm as Cl_2) continuously in potable water systems, not to exceed under any circumstances the 4 mg/l limit for drinking water in the USA (CFR 2002). Local and regional regulations should be consulted because in some countries, secondary disinfection of potable water with chlorine is not allowed.

Precautions regarding performance of chlorine based disinfectants are that care must be taken on-site to safely handle the chemical or the gas (unless halide oxidation by electrolysis is used), pH above 8.0 severely reduces antimicrobial performance, heavy deposits such as biofilms are not readily penetrated by typical use concentrations, the *Legionella* host forms cysts that are highly resistant to chlorine, chlorine demand from organic and inorganic chemicals in the system can consume the disinfectant, hot water systems are notoriously hard to treat with chlorine because the oxidant is consumed very quickly and chlorine can be very corrosive.

In difficult-to-control systems, hot water disinfection with copper–silver ionization (see Section 7.3.1.1) and coldwater disinfection with chlorine or mixed oxidants is to be highly recommended.

7.1.6.3 Chlorine dioxide

Chlorine dioxide is a unique oxidizing agent. It does not react with water (as does chlorine to form hypochlorous acid) and is therefore a dissolved gas at use concentrations. Chlorine dioxide is powerfully antimicrobial at low concentrations (sub-parts per million) yet it does not chlorinate many of the types of organics that cause loss of viability in microorganisms nor does it react with the organics in water that cause the most concern about toxic by-product formation (chlorination by-products are haloforms such as trihalomethanes (THMs), haloacetic acids (HAA) such as chloroacetic acid and haloketones). Chlorine dioxide uniquely oxidizes inorganic compounds compared to chlorine. For example, chlorine water applications will generate colloidal sulfur in water treatment if there are any sulfides (reduced sulfur) present; this can cause many problems including corrosion, blockage and odor/appearance complaints. At use concentrations, chlorine dioxide rapidly oxidizes sulfide to sulfate (not elemental sulfur) which is generally innocuous in water.

Secondary disinfection applications of chlorine dioxide are effective in laboratory simulations of real-world domestic water systems (Thomas *et al.* 2004; Loret *et al.* 2005) and in field use for extended periods (Walker *et al.* 1998; Srinivasan *et al.* 2003) at much lower oxidant concentrations (typically less than half) compared to effective oxidant concentrations from chlorine or chloramines.

Since chlorine dioxide is not stable enough to ship, on-site generation is necessary. Several precautions are required because the gas is explosive in percentage concentrations and because unreacted precursors (chlorite and chlorate) are tightly regulated due to their significant toxicities. There are many approaches to on-site generation of chlorine dioxide including electrolytic processes that require feeding only one reactant (http://www.haloxtech.com) and multiple precursor systems (http://idiclo2.com/clo2chem/safety.html). Given in Figure 7.6 is the schematic of a

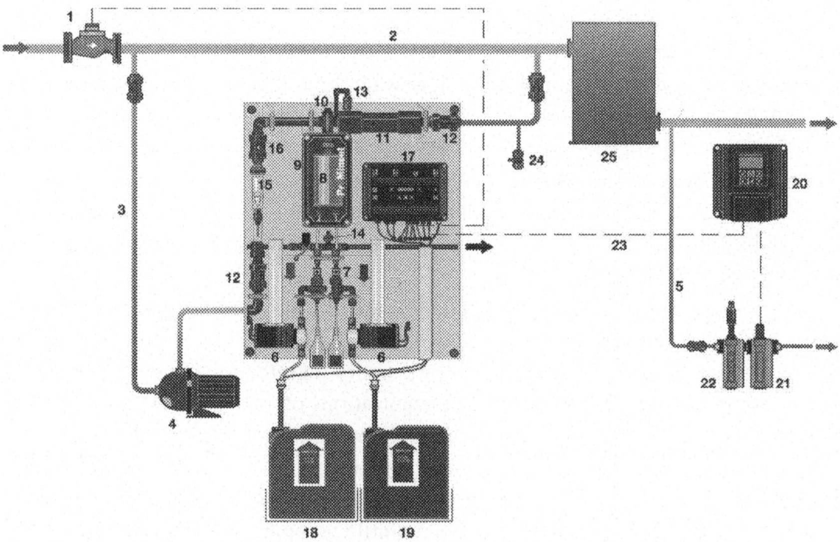

Figure 7.6 ProMinent BelloZon® Chlorine Dioxide Generation is an attractive alternative to chlorine disinfection because chlorine dioxide produces no toxic or potentially toxic organic chlorine compounds such as chlorophenols or halogenated hydrocarbons or haloforms such as chloroform. Chlorine dioxide can also be more effectively used to disinfect high pH, low hardness water compared to chlorine. (1) water gauge (contact, analog), (2) main water pipe, (3) bypass pipe 1–2 m^3/h, (4) bypass pump, (5) water monitoring pipe, (6) solenoid dosing pump, (7) flow sensor, (8) reactor, (9) reactor housing, (10) dosing valve (pressure sensitive), (11) mixer, (12) stop valve, (13) ventilation valve, (14) suction device, (15) bypass monitor, (16) non-return-valve, (17) control with production level display, (18) acid in safety vessel, (19) chlorite in safety vessel, (20) D1C – chlorine dioxide meter, (21) chlorine dioxide probe, (22) water readings monitor, (23) interlocking contact, (24) rinsing connector, (25) holding tank (reaction time 10–15 min). Artwork provided by ProMinent Fluid Controls, Inc.

widely applied device that has successfully stood the test of time and thousands of field installations.

Treatment parameters for chlorine dioxide are typically set at 0.4–0.6 mg/l (ppm as ClO_2) continuously in potable water systems, not to exceed under any circumstances the 0.8 mg/l limit for drinking water in the USA (CFR 2002).

Precautions regarding performance of chlorine dioxide are that care must be taken to safely handle precursor chemicals, chlorine dioxide by-products are strictly regulated in drinking water, chlorine dioxide is very volatile (it is a dissolved gas in water), it has approximately the same toxicity as gas chlorine with an even lower exposure limit and it is corrosive at high concentrations. (http://www.cdc.gov/niosh/pel88/10049-04.html).

7.1.6.4 Chloramines

When chloramines are used as the primary disinfectant by municipalities in the United States, there has been apparently less legionellosis associated with those systems compared to systems in which chlorine has been used by the municipality as the primary disinfectant (Kool *et al.* 1999; Heffelfinger *et al.* 2003; Stevens *et al. in press*, Effect of municipal water disinfection with chloramines on *Legionella* colonization of potable water systems: a prospective study. Poster Q-183. *105th Annual Meeting of the American Society of Microbiology*, Atlanta, Georgia 2005).

Combined chlorine residual oxidant appears to be very effective in control of Legionella and biofilms in laboratory experiments (Stewart *et al.* 2001; Storey *et al.* 2004c; Loret *et al.* 2005).

At present, about 25% of the municipalities in the United States use monochloramine as the primary disinfectant (Kool *et al.* 1999). A municipality may convert to chloramines from chlorine for primary disinfection because of concern about chlorination by-products. The concentrations of THMs and HAA chlorination by-products in drinking water are regulated in the USA (CFR 2002).

Unfortunately at present, there is no practical and safe means to provide monochloramine for secondary disinfection. This is because concentrated solutions of chloramines are too unstable, corrosive, volatile compared to chlorine water (i.e., hypochlorous acid) and are more irritating to skin and eyes.

7.1.6.5 Ozone

Ozone is effective against *Legionella* but is not recommended for secondary disinfection in building water systems. The secondary disinfection problem with ozone is that chemical reactivity is so great as to make it impossible to hold an antimicrobial residual throughout the system which leaves most of it unprotected and subject to microbial fouling (EPA 1999; Thomas *et al.* 2004; Loret *et al.* 2005). Disinfection with ozone can be effective in certain specialized applications.

7.1.6.6 UV irradiation

Ultraviolet irradiation at 254 nm wavelength is useful for control of biological hazards (Eckmanns *et al.* 2002; Hall *et al.* 2003) but because no residual antimicrobial activity is imparted to the treated water, it should be combined with other methods of disinfection. The efficacy of the ultraviolet irradiation on the hot water system of a hospital colonized by *L. pneumophila* was effective in a relatively small area even when the incoming water was highly contaminated but other methods of disinfection are necessary to control microorganisms through the water system (Franzin *et al.* 2002). Maintenance of the UV system is necessary in order to prevent biofilm formation on the quartz sleeve of the lamp.

7.1.6.7 Other chemical disinfectants

Other control methods for water systems in dental offices and other healthcare facilities have been recently evaluated including iodine, enzymes (proteinases and carbohydrases), surfactants, peroxyacids (such as hydrogen peroxide, peracetic acid, butylhydroperoxide, and combinations of these) and hydrogen peroxide/silver

(Shetty *et al.* 1999; Zinkevich *et al.* 2000; Walker 2003). These treatments are variously effective and should always be used in conjunction with one or more of the preferred methods of control in domestic water systems and in accord with local regulations.

7.2 HAZARD CONTROL IN UTILITY WATER SYSTEMS

Although most legionellosis results from exposure of people to contaminated potable water systems, the most well-known outbreaks are associated with utility water systems in buildings.

7.2.1 Cooling water systems

In nearly every utility water system, cooling towers should be critical control points (see Section 6.3) for preventing legionellosis.

7.2.1.1 Function

The function of water cooling towers is to transfer waste heat from a building or process into the atmosphere through evaporation. This process is efficient because the heat capacity of water is very high and because a tremendous amount of heat transfers to air when water evaporates. Figures 7.7–7.9 illustrate typical cooling tower configurations.

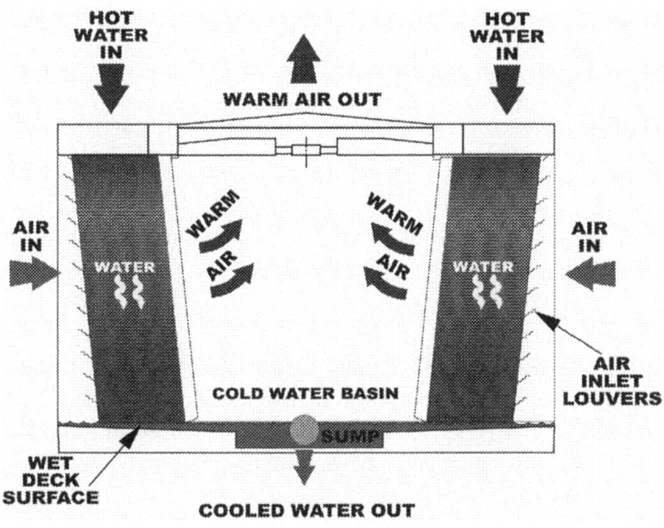

Figure 7.7 Typical crossflow cooling tower of induced draft, double flow arrangement. Hot water enters at the distribution basin on top of the unit, is sprayed over the wet deck material where it is cooled before falling into the sump at the bottom of the unit. From here the cooled water is piped to the heat load. Cool air enters on the unit side, directly contacts the water as it passes through the wet deck and then exits as warm, humid air at the top of the unit. Artwork courtesy of Baltimore Aircoil Company, Baltimore, Maryland (www.baltimoreaircoil.com).

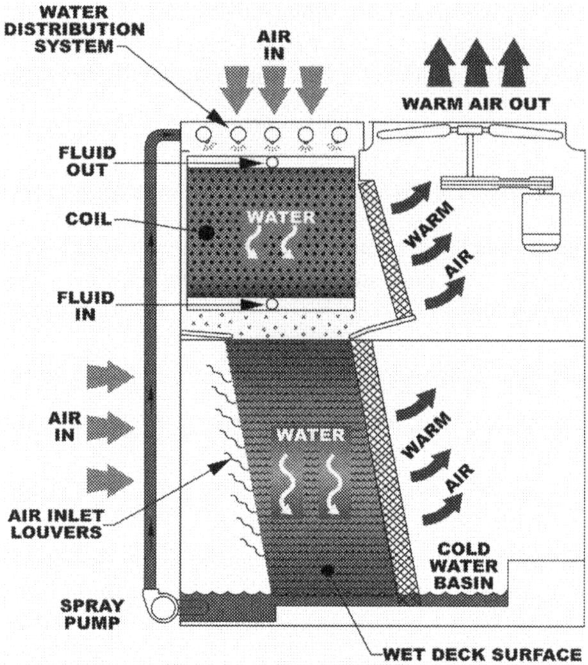

Figure 7.8 Typical combination open and closed circuit cooling tower. The coil of the closed circuit cooling tower portion is located at the top of the unit and the wet deck of the open circuit cooling tower portion is located at the bottom of the unit. Cooled water from the sump is sprayed over the coil at the top left of the unit and after flowing over the coil (where it is heated) flows over the wet deck (where it is cooled) and falls into the unit sump where it is pumped to the top of the coil to repeat the process. Cool air is drawn in the top of the unit and flows downward over the coil. Cool air is also drawn in the side of the unit and flows through the wet deck. The two, now warm and humid air streams then mix and exit at the right top on the unit. Hot water (separate from the spray water) enters the coil at the bottom and exits as cooled water at the top of the coil. From here, this cooled water is piped to the heat load. Artwork courtesy of Baltimore Aircoil Company, Baltimore, Maryland (www.baltimoreaircoil.com).

All of these systems use fans to move air through recirculated water which is usually within the temperature range for *Legionella* growth (20–50°C, 68–122°F). From the cooling water process, water vapor is introduced into the atmosphere. Drift eliminators are designed to limit dissemination of water aerosol to the surrounding environment.

7.2.1.2 Control recommendations for cooling towers

Cooling water systems should be kept clean of scale, corrosion, debris, and microbial fouling. An effective microbial control treatment should be used to eliminate, prevent or reduce biological hazards to acceptable levels. Even in well-maintained cooling water systems, periodic cleaning and disinfection should be implemented.

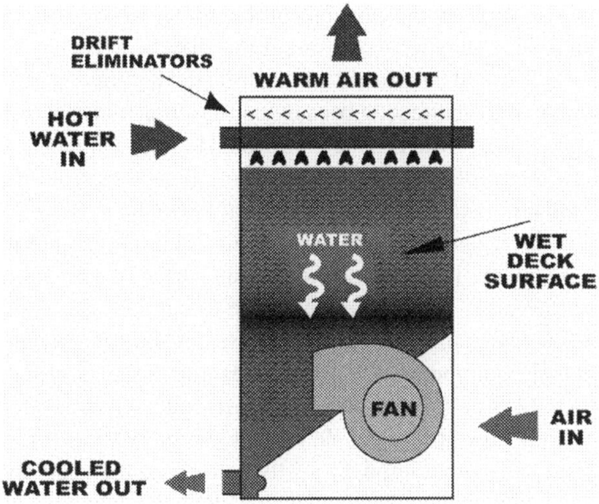

Figure 7.9 Typical forced draft counterflow cooling tower. Hot water enters the pressurized distribution system at the top of the unit and is sprayed over the wet deck material. The cooled water falls into the pan at the bottom of the unit. From here it is piped to the heat load. Cool air enters on the unit side and is blown up through the wet deck to exit as warm, humid air at the top of the unit. Artwork courtesy of Baltimore Aircoil Company, Baltimore, Maryland (www.altimoreaircoil.com).

It is recommended that the services of a qualified water treatment specialist should be used to oversee the water treatment program.

Note that the following general recommendations do not supersede local regulatory guidance or regulations. First establish a hazard analysis and critical control plan (Chapter 6) to properly manage implementation of hazard control.

Hazard control in cooling water systems should be comprised of procedures to (1) periodically inspect and maintain the system, (2) clean and disinfect the system, (3) pre-condition the makeup water, (4) control scale, (5) control corrosion, and (6) control microbial fouling.

- *Inspection and maintenance.* Periodic visual inspection and regularly scheduled maintenance of the system are essential for hazard control. Special attention should be devoted to making sure drift eliminators are in good condition. High efficiency drift eliminators should be used; older cooling towers should be retrofitted. Regular maintenance is necessary both to control *Legionella* growth and for effective operation. The system should be properly monitored and maintained to prevent buildup of scale and sediment and bio-fouling, all of which will support *Legionella* growth and reduce operating efficiency.
- *Cleaning and disinfection.* Even well-maintained cooling towers should be cleaned and disinfected at least twice a year. Normally this maintenance will be performed before initial start-up at the beginning of the cooling season and after shutdown in the fall. Continuously operated systems can be cleaned and disinfected on-line or by intermittently shutting of the fans for short periods

(a few hours). Systems with heavy bio-fouling or high levels of *Legionella* may require additional cleaning. Any system that has been out of service for an extended period should be cleaned and disinfected before restart. New systems require cleaning and disinfecting because construction material residue can contribute to *Legionella* growth. Disinfection should be with an initial shock treatment of 5–10 ppm free *total* residual chlorine using bleach or from electrolytically generated oxidant on-site. Recirculate the water *with the fans off* for at least 5 h. Use of detergent or dispersant to disperse bio-fouling is recommended. If the system can be taken off-line, physically clean surfaces. Take care to not damage cooling tower fill. There is no evidence to support a recommendation to remove cooling tower fill (packing) for physical cleaning. *Caution*: Wear protective personal equipment (PPE) when cleaning surfaces especially if high-pressure washing is performed. PPE should include at least an N-95 mask and take care to limit exposure of people in the vicinity to water spray or debris. A water-resistant suit with a hood, protective gloves, and a properly fitted respirator with a high-efficiency particulate (HEPA) filter or a filter effective at removing 1 μm particles is recommended for cleaning personnel. After cleaning, maintain 5 ppm total residual chlorine in the system with recirculated water for 24 h. Repeat the cycle until there is no visual evidence of biofilms in the system. *Caution*: Chlorine is corrosive. Exceeding these recommendations could cause severe damage to the cooling tower and its components.

- *Water pre-conditioning*. The quality of water used in cooling systems can often be substantially improved by simple filtration and softening. This processing can reduce the quantity of water treatment chemicals needed to control scale, corrosion and microbial fouling. Water pre-conditioning should be considered for every cooling water system. Generally, pre-conditioning equipment for make-up water to cooling systems is grossly underutilized especially in the USA.
- *Scale control.* Scale is the deposition of inorganic salts, such as calcium carbonate, onto surfaces. Its formation is influenced by alkalinity, pH, temperature, concentrations of calcium and magnesium and total dissolved solids in the water. Depending on water quality and operating conditions, scaling can be controlled with chemical inhibitors such as phosphate, phosphonates and polymers. The natural concentrations of phosphate and phosphonates in aquatic environments are very low. Therefore, use of these chemicals in cooling water fertilizes growth of microorganisms in these systems. Excessive use of scale inhibitors to control scaling can result in massive microbial fouling. In many cases, the use of scale inhibiting chemicals can be reduced or eliminated by removing calcium and magnesium hardness from makeup water by ion-exchange softening.
- *Corrosion control.* Most cooling water systems are comprised of metals such as mild steel which are highly susceptible to corrosion. Even corrosion-resistant alloys such as stainless steel or copper alloys are subject to corrosion in cooling water systems. The buildup of corrosion products on surfaces contributes to conditions which allow for proliferation of microorganisms in the system. Corrosion can be inhibited by use of chemical inhibitors such as phosphates, azoles, molybdenum and zinc. Water conditions such as pH and chloride

concentration substantially influences corrosion rates. Water pre-conditioning and pH control can help minimize corrosion in cooling water systems.

- *Microbial control.* Even in cooling water systems where water pre-conditioning has been used to minimize scale and corrosion, microbial control is required. Two classes of industrial biocides are used for this purpose: Oxidizing biocides and non-oxidizing biocides. Typically, an oxidizing biocide such as chlorine, bromine, stabilized bromine, chlorine dioxide, or ozone should be used continuously. In addition for most systems, intermittent shot doses of non-oxidizing biocide two or three times a week is usually necessary to control microbial fouling in cooling water systems. The preferred non-oxidizing biocides used for cooling water microbial control include glutaraldehyde, isothiazolone, and dibromonitrilopropionamide but there are many others to choose from. These products must be used according to government-regulated label instructions and should be applied under the guidance of a qualified water treatment professional. Since the microbial control program is critical to control of *Legionella* in cooling water systems, more detail of available treatments is given as follows.

7.2.1.3 Microbial control in cooling towers

Industrial water systems are never sterilized as a result of microbial control programs applied. Even if enough chemical or other remedy could be added to achieve sterilization, the system would rapidly become re-inoculated with microorganisms since cooling systems are open to the environment. The most significant practical consequence of this fact is that increasingly tolerant microbial communities comprised of the survivors of the applied antimicrobial treatment are selected for especially in diverse microbial biofilms (Srikanth and Berk 1994; Sutherland and Berk 1996; Stewart *et al.* 2001). It is advisable therefore to proactively vary the antimicrobial stresses in the cooing water microbial control program (McCoy 2003b). For example, a practical way to vary antimicrobial effects is to alternate between continuous-dosing and slug-dosing oxidizing antimicrobial chemical. Another effective strategy to control microbial fouling is to alternate use of an oxidizing antimicrobial with a non-oxidizing antimicrobial to ensure that antimicrobials with different modes of action are applied. Vary the treatment protocol based on observed antimicrobial performance in the system. Table 7.1 gives a hypothetical example of a microbial control program designed to proactively vary antimicrobial stresses in a cooling water system.

Antimicrobial control methods for *Legionella* in water systems have been extensively reviewed (Kim *et al.* 2002; WHO 2002; Thomas *et al.* 2004; Loret *et al.* 2005). For industrial water systems, chlorine, bromine, stabilized bromine, stabilized chlorine and chlorine dioxide are the preferred oxidants used for microbial control. Also used effectively for industrial water systems are the "non-oxidizing" antimicrobials. Especially effective active ingredients are dibromonitrilopropionamide, glutaraldehyde, and isothiazolones, but there are many others to choose from (Kim *et al.* 2002). These products are effective when used properly. As has been previously discussed, control of the pathogen host (the protozoa) is critical in legionellosis prevention. Currently, the best practical means is to control the

Table 7.1 An example of variation in antimicrobial treatments to control microbial fouling in an industrial cooling water system.

Protocol*	Oxidizing biocide	Non-oxidizing biocide
A	Continuous oxidant controlled to 0.5–1.0 mg/l total residual as Cl_2	2 mg/l a.i. isothiazolone shot dose 3x per week
B	Same as A with daily shot doses of oxidant to 1–2 ppm for 1 h per day	40 ppm a.i. glutaraldehyde shot dose 3 times per week

* Alternate between A and B when heterotrophic plate counts exceed predetermined limit (e.g., $>10^5$ CFU/ml or if there are visible signs that microbial fouling has increased).

biofilms upon which protozoa graze (Donlan 2002; Donlan *et al.* 2002). This can be very difficult especially in dynamic systems.

7.2.1.4 Emergency disinfection of cooling towers

Emergency disinfection is costly and potentially hazardous. It should be therefore avoided. The best way to avoid emergency disinfection of cooling towers is to establish and implement effective water safety plans (Chapter 6) and establish hazard control procedures described in Section 7.2.1.2.

The general recommendations given here do not supersede recommendations given by local or regional governments. See Table 5.1 for listing of government-approved guidance in your area and before proceeding. In the USA, consult guidance given by Occupational Safety and Heath Administration (OSHA 1998).

- Take the cooling water system off-line. Shut off fans. Limit access to the area. Clean and disinfect the entire cooling system with 20 ppm *free residual* chlorine bleach or from electrolytically generated oxidant on-site. Add dispersant to cooling water system. Recirculate water *with the fans off* for 1 h. Drain the system. Refill the system. Repeat disinfection with 20 ppm *free residual* chlorine bleach or from electrolytically generated oxidant on-site. Drain the system. Refill and maintain 10 ppm *total* residual chlorine for 24 h. Drain system. Inspect system for visual evidence of biofilm. If found, repeat emergency disinfection. Use PPE (see above) for cleaning and disinfection operations.
- Refill with clean water in accordance with an effective water treatment program. The unit should be now ready for service. Document operation and maintenance in a log or maintenance records book.

Caution: Chlorine is corrosive and will shorten the service life of metal plumbing and cooling tower components. Avoid the need for emergency disinfection.

7.3 HAZARD CONTROL IN POTTING SOIL

Potting soil has been confirmed as a potential source of infective particles that when inhaled into the lungs can cause legionellosis in susceptible people. The following precautions to control the hazard should be also taken when handling potting mix, peat, mulch, and garden compost.

- Before fully opening the potting soil bags, make a small opening in the bag to allow inserting a garden hose or water-can spout.
- Moisten the potting soil.
- Open potting soil bags with care to avoid inhaling airborne particles.
- Wear gloves when handling potting soil to avoid transferring from hand to mouth.
- Wash hands thoroughly after handling potting soil even if gloves have been worn.

8

Validation and verification

Validation is quantitative evidence (data) that hazard control methods have actually worked under operating conditions to eliminate, prevent or reduce hazards to acceptable levels. But exactly what constitutes evidence that the hazard has been eliminated, prevented, or reduced to safe levels? Since the "infectious dose" of *Legionella* is not known and anyway, it varies depending on the susceptibility of the victim, therefore any amount of detectable viable pathogen is potentially dangerous to someone.

The recommended safe level of *Legionella* in water systems varies by country (see Section 8.1.1.3). In the author's opinion, it is both practical and consistent with the science to set zero detectable *Legionella* (by the standard method with no concentration step, Section 8.1.1.1) as the level for which hazard control is said to have been validated under operating conditions. This opinion should not be construed to supersede local or regional regulations.

Verification is the periodic auditing process to document that the validated plan is being implemented properly. For most building water systems, verification should be done by an independent auditor at least once a year or when major changes to the system have occurred.

8.1 VALIDATION

In the USA, the National Primary Drinking Water Regulations (NPDWRs or primary standards) are legally enforceable standards that apply to public water systems.

Primary standards protect public health by limiting the levels of contaminants in drinking water. The maximum containment limit goal (MCLG) is zero detectable *Legionella* by treatment technique in the primary drinking water standard (CFR 2002, see also Section 5.2.5). The United States Environmental Protection Agency (USEPA) believes that effective treatment to control *Giardia*, coliforms, and viruses (which all have zero detectable MCLGs) will also control *Legionella* in drinking water. The limit for heterotrophic plate counts (HPC) in drinking water is 500 CFU/ml. The heterotrophic bacteria enumerated with the HPC cause no known health effects; the HPC is an analytic method used to measure a wide variety of bacteria common in potable water. The lower the concentration of bacteria in potable water, the better maintained is the water system.

For hazard analysis and control (Chapter 6), validation that the hazard has been eliminated, prevented or reduced to acceptable levels under operating conditions is required. Scientifically proven analytic methods to quantitatively measure the hazard at the critical control points (CCPs) and throughout the facility are used to validate that control of the hazard has been achieved. Environmental samples of water at the CCPs and distal sites should be taken periodically for microbiological analyses such as total heterotrophic bacteria concentration and *Legionella* concentration. The frequency of this testing should be left up to the discretion of the hazard analysis and critical control point (HACCP) or water safety plan (WSP) team. In some systems, it is well enough to initially validate that the plan is adequate for control of the hazard and then check periodically, say once or twice a year or once a quarter. For the prevention of legionellosis, this is the most cost-effective and scientifically defensible way to use data from *Legionella* tests.

8.1.1 Quantitative analysis of the hazard

A commonly misapplied activity in legionellosis control is analysis to determine pathogen concentrations in samples from engineered water systems. In the worst cases, pathogen analytic results are used wastefully, counter-productively and therefore, dangerously.

The essential problem statement is this: *Legionella* concentrations in water systems cannot been correlated to quantitative determinations of legionellosis risk. Negative *Legionella* test results from water systems are not necessarily equivalent to a low risk of contracting legionellosis. Positive *Legionella* test results cannot be used alone to determine the risk of legionellosis.

Most scientists agree that it is not currently possible to determine the risk of legionellosis from measurements of *Legionella* concentrations because:

1. The infective dose of pathogen is not known.
2. The infective dose varies according to the susceptibility of the exposed population.
3. The test method requires 10 days and is therefore only a look backwards to what the pathogen concentration might have been when sample analysis began.
4. Effects due to sample transit *en route* to the laboratory are not controlled.
5. The risk of legionellosis also depends on the risk of population exposure and inhalation of aerosols containing *Legionella*.

However, when results are properly interpreted within the context of a risk management plan, pathogen analytic services can be very helpful. Pathogen analysis of environmental samples should be used to confirm that control measures have been effectively applied to reduce the hazard to acceptable levels, eliminate it or prevent it from harming people under operating conditions. Confirmation that control measures are working (validation) is a necessary part of every hazard analysis and control plan. Also, results can be used diagnostically to identify hazards in contaminated or infested areas in a water system.

8.1.1.1 The standard culture method

The first isolations of *Legionellae* were from guinea pigs and from embryonated chicken eggs (McDade *et al.* 1977) and on culture media with Mueller–Hinton agar supplemented with hemoglobin and L-cysteine (Feeley *et al.* 1978). The plate culture method was refined and developed into the Feeley–Gorman medium. By this innovation, quantitative analysis of viable *Legionella* in all sorts of environments was made possible. Starch was replaced with charcoal to detoxify the medium and the amino acid source was changed to yeast extract. This led to use of the medium known as buffered charcoal yeast extract (BCYE) and is today the basis for all analytic culture methods.

Several refinements and improvements resulted in the currently preferred BCYE medium that is enriched with α-ketoglutarate (Edelstein BCYE-α medium) with and without selective antimicrobial agents and indicator dyes. In order to isolate certain species, this medium can be supplemented with bovine serum albumin (Fields *et al.* 2002).

The standard method ISO 11731 has been developed (ISO 1998) which specifies use of the BCYE-α medium supplemented with ammonia free *g*lycine, *v*ancomycin, *p*olymyxin B, and *c*ycloheximide (GVPC). In addition to these supplements, GVPC contains ferric pyrophosphate, L-cysteine, α-ketoglutarate. This method is generally consistent with the original method developed by the Centers for Disease Control and Prevention (B. Fields, personal communication) and with standard methods used in Australia and Singapore (AU/NZ 3896). A method that is substantially similar to these is used in France (AFNOR T90-431).

Selectivity steps are required to inhibit competition from faster growing bacteria that may overwhelm *Legionella* in the sample. Samples are split in three; one is plated with no treatment, one sample is heated (30 min at 50°C), and the third is acidified with HCl (pH 2.2 for 5 min). Concentration of the sample by centrifugation (6000*g* for 10 min) or membrane filtration is optional depending on detection limit requirements.

Standard culture methods specify pathogen confirmation steps that require comparative growth on BCYE-α agar and blood agar (or nutrient agar) not supplemented with iron and L-cysteine. Colonies that grow on GVPC but fail to grow on BCYE without L-cysteine and blood (or nutrient) agar are confirmed *Legionella* spp.

Serogroup typing is then performed on confirmed *Legionella* colonies by latex bead agglutination to differentiate serogroup 1 from serogroup 2–14 and from other serogroups. With the suitable range of antisera available, typing and subtyping to full species and sub species is possible.

Several factors limit the culture method. Analyst experience with the method directly correlates with pathogen recovery and higher concentrations. The method requires 10 days to yield confirmed results. Preparation of the medium is tricky and requires expensive quality control. The pathogen is sensitive to factors that are difficult to control during sample transit. Concentration steps used to achieve lower detection limits are notoriously inefficient and irreproducible (<50% of viable *Legionella* are recovered during sample concentration processing and variation among laboratories is high).

8.1.1.2 Molecular methods

Molecular methods are faster, less expensive, less subjective, more sensitive, and can be performed in the field. They all suffer a critical limitation: none of the commercially available molecular methods can differentiate viable *Legionella* among the background of nonviable bacteria in environmental samples (however, real-time polymerase chain reaction (PCR) methods show promise). Thus, in practice only the culture method can detect the effect of disinfection and this is an essential requirement for a pathogen analytic method to confirm effective control in engineered water systems.

Molecular methods include nucleic acid detection using the PCR or fluorescence *in situ* hybridization (FISH), and serologic methods by antigen/antibody reactions detected with enzyme linked immuno-specific assays (ELISA) or differential fluorescent antibody direct cell counting (Cloud *et al.* 2000; Fields *et al.* 2002; Campos *et al.* 2003). Most scientists agree that these detection systems are useful in the clinical laboratory for diagnosis and serogrouping. In fact, dependence on the culture method in the clinical laboratory has been substantially reduced because the molecular methods are superior in so many ways. Advances in these methods have improved clinical diagnosis of legionellosis. Quantitative differentiation of viable *Legionella* is not a requirement in most clinical applications.

For environmental or industrial samples, nucleic acid or serologic methods should be used only as a rapid screen to identify those samples free of any *Legionella*, whatsoever. This service can be helpful in some risk reduction situations. However unfortunately, this service does not eliminate the need for culturing samples to quantify viable *Legionella* in samples from water systems.

8.1.1.3 Using analysis of pathogen to validate hazard control

Given the ability to measure pathogen concentrations in water systems, a natural tendency then is for people to demand action (or "trigger") levels based on these analyses. The overwhelming consensus of scientific opinion is that such trigger levels are entirely arbitrary. They are not established on any scientifically relevant or experimentally verifiable basis. They are however, still useful.

The justification for having trigger levels correlated to corrective action is that they provide a numeric framework for practice in the field, they can be used to validate the implementation of control methods and they are a basis for expedient

Table 8.1 Trigger levels given in guidance documents for *Legionella* control in cooling water.

Legionella concentration (CFU/l or CFU/ml)	Corrective actions required
≤100 CFU/l ≤1000 CFU/l* *≤100 CFU/ml*	Acceptable control. No remedial action required.
100 to 1000 CFU/l 1000 to 100,000 CFU/l* *100 to 1000 CFU/ml*	Immediately review treatment program and develop remedial risk management plan. Confirm results immediately by re-sampling.
≥1000 CFU/l ≥100,000 CFU/l* *≥1000 CFU/ml*	Immediately implement corrective action by shot-dose antimicrobial. Clean and disinfection required. Immediately resample. Review risk management plan and control program. Reassess the system. In France, immediately shut down the facility (see Section 5.2.4.1).

UK and much of Europe; *France; *US, AU and Singapore*; *Note: French regulation effective July 1, 2005 requires shut down of the facility if counts exceed 100,000 CFU/L.

communications. This is helpful. Table 8.1 gives a compilation of corrective action levels from various guidance documents for cooling tower systems. Corrective actions specified are generally the same even though the trigger levels vary as much as three orders of magnitude in codes of practice from around the world. The tremendous differences in these trigger concentrations are at first disconcerting and then confusing. But since the numeric values of trigger levels are arbitrary anyway, the practical consequence of differences is irrelevant so long as trigger concentrations and the actions required are applied consistently in a given region.

These limits are based on what was the perceived detection limits for the assay at the time the limits were established. Thus, the UK action levels (which are three orders of magnitude lower than those specified in AU, Singapore, US) are based in part on the notion that a concentration of the sample by centrifugation could achieve such a low detection limit. However, the concentration step is controversial because recovery of *Legionella* from centrifuged environmental samples is low (less than 50% and sometimes not even 30% recovered). Similarly, another limit is based on sample concentration by membrane filtration that also fails to reliably recover the pathogen from environmental samples. In regions where the concentration step is not required, the detection limit of the assay is much higher and so are the action levels.

Corrective action levels for domestic water service systems, such as potable water for buildings, are also specified in guidance documents and they are also arbitrary. But they are similarly linked to consistent action recommendations, regardless the absolute value of the trigger level, and therefore they serve a useful practical purpose. Table 8.2 gives the action levels published by the Occupational Safety and Health Administration (OSHA) in the USA (OSHA 1998).

One rational suggestion to improve this situation is that there should be a global consensus on arbitrary trigger levels for corrective actions based on a scientifically defensible detection limit for *Legionella* in the standard culture method. This

Table 8.2 Action levels for water systems in the US (OSHA 1998).

	Legionella detected with the standard method (CFU/ml)		
	Cooling water	Potable water	Humidifiers
Action level 1	100	10	1
Action level 2	1000	100	10

Action level 1 = promptly clean system and/or use biocide or disinfectant, if applicable.
Action level 2 = immediately clean and disinfect the system. Take immediate steps to limit employee exposure.

would reduce confusion, especially for international organizations that seek to apply consistent criteria for validation that the *Legionella* hazard has been controlled in their facilities, worldwide. Due to the uncertainties that arise when environmental samples are concentrated by centrifugation or membrane filtration, it would be best to not perform these operations, in the author's opinion. If that technical point were conceded, then the higher trigger levels would stand as the most defensible. Since there would be no need to concentrate samples, the ISO 11731 method (without a concentration requirement) could be recommended, worldwide.

8.1.1.4 Examples of hazard control validation

An excellent example of a practical approach to using data from *Legionella* tests for validation of hazard control is given by the Allegheny County Health Department. In its guidance "Approaches to the Prevention and Control Of *Legionella* In Allegheny County Health Care Facilities" (PA 1997), secondary disinfection of the potable water system is recommended if the percentage of positive cultures (using the standard method) indicate viable *Legionella* is present in more than 30% of the distal sites sampled. Note that the validation criterion is not viable cell concentration (such as is required in many codes of practice) but rather, it (the validation criterion) is the percentage positive indications using the standard culture method. If the percentage of positive viable cultures is less than 30% of the distal sites sampled, then secondary disinfection of the potable water system is not recommended until the problem has been clarified as to the source of the contamination. This guidance recommends that every site should be surveyed at least once per year. In hospitals with 500 beds or less, 10 distal sites should be sampled. In hospitals with greater than 500 beds, two distal sites per 100 beds should be sampled. The distal sites should be selected from units housing patients at high risk of susceptibility to legionellosis (e.g. transplant or chemotherapy units). The efficacy of these recommendations has recently been quantitatively confirmed (Squier *et al.* 2005) by careful measurements of nosocomial case rates before and after implementation. The data show that implementation of hazard analysis and control has significantly prevented nosocomial legionellosis (see Section 4.2.1 for details).

Another example of validation criteria comes from specifications given in the new French regulations which became enforceable July 1, 2005 (France 2005a–c). Open cooling tower systems with thermal capacities greater than 2000 MW must be authorized (permitted) and all other cooling water systems, open or closed,

Table 8.3 Validation criteria based on French regulations for cooling towers (France 2005a–c).

Product: utility water	CCP (number)	Validation criteria	Frequency
Processing step			
Cooling			
Cooling towers with thermal capacity >2000 MW	1B	Legionella control validation: less than 1000 CFU/l viable *Legionella* by the standard method AFNOR T90-431	Monthly
Cooling towers with thermal capacity <2000 MW		Legionella control validation: Less than 1000 CFU/l viable *Legionella* by the standard method AFNOR T90-431	Bi-monthly
Exceptions		Facilities in which the results of analyses are less than 1000 CFU/l during 12 monthly (facilities with permits) or 6 bi-monthly (registered facilities), consecutively	Quarterly

must be registered in France. Periodic environmental monitoring (surveillance) is required to determine the concentration of viable *Legionella* bacteria in the water of all cooling towers. Data from this activity is necessary to quantitatively document that control of the hazard has been achieved under operating conditions (validation). The French standard plate count method for *Legionella* (AFNOR T90-431, see Section 8.1.1.1) is to be used by laboratories that have been approved (accredited) by the government. The environmental surveillance schedule to validate hazard control under operating conditions is specified in the regulation: monthly *Legionella* analyses for facilities subject to authorization (permits), bi-monthly for facilities subject to registration, and tri-monthly (quarterly) for facilities in which the results of the analysis less than 1000 CFU/l during 12 consecutive months (Table 8.3).

If the results of quantitative viable *Legionella* analysis in cooling water indicate that concentrations are greater than or equal to 1000 but less than 100,000 CFU/l, then the system must be immediately cleaned and disinfected; a new analysis to determine the viable *Legionella* concentration must be performed within at least 15 days; if concentrations remain in this range (1000–100,000 CFU/l), then a new hazard analysis and control plan with improvements must be established and repeated until the concentration is below 1000 CFU/l.

If the concentration of viable *Legionella* in the cooling water is shown to be greater than 100,000 CFU/l, then the facility owner/manager must immediately inform the local government authority and immediately shutdown the facility for cleaning and disinfection according to the corrective action previously established. The owner/manager of the facility must place a notice for public view on the cooling tower which reads (paraphrased) “Urgent and Important – Cooling Tower Exceeds Limit of 100,000 CFU/l *Legionella*”. Before the facility can be restarted, the owner/manager must establish a new hazard analysis and control plan with improvements in both hazard control methods and verification. Environmental samples must be analyzed by the standard plate count method 48 h after restart of the facility and then at least every 15 days thereafter for 3 months. If during this period the concentration of *Legionella* is shown to be above 10,000 CFU/l, then the facility is to immediately shut down again, report to the local government authority and repeat the corrective action procedure before restarting the facility. In cases where shutdown of a facility poses grave risk, exceptions to the shutdown rules can be made by special permission from the government, but more frequent cleaning and disinfection and more frequent environmental surveillance of viable *Legionella* analysis using the standard method will be required (see also Section 5.2.4.1).

8.1.1.5 Sampling protocol

General guidance on sampling protocol is given as follows. These recommendations do not supersede local or regional guidance for sampling and shipping specimens for *Legionella* analysis.

Water sample collection
Take about 1000ml samples of water from cooling towers, condensers, storage tanks, drinking water faucets (immediate and post 30 s flush), sink faucets, showers (immediate and post 30 s flush), pools, spas, hot water tanks (immediate and post 30 s flush) and ice from ice machines using pre-sterilized hinged-lid containers with thiosulfate neutralizing tablet (e.g. https://www.hardydiagnostics.com/catalog2/user/ProductDetails.asp?poid=EPA120T83P).

Deposit sample collection
Use pre-sterilized thiosulfate-neutralized swab kit (e.g. EnviroTrans, https://www.hardydiagnostics.com/catalog2/user/ ProductDetails.asp?poid=SRK15) to swab (4 times) surfaces from cooling towers, condensers, storage tanks, ice machines and, after moistening the outlet with a trickle of water, the following outlets: water faucets, sink faucets, showerheads, and hot water tank faucets.

Shipping requirements
Use shipping labels on watertight insulated shipping containers. Clearly indicate sample location, date and time. Next day shipping to the laboratory is required. Ship within 8 h of sample collection and do not allow samples to freeze nor excessively warm. Do not send samples to the laboratory on Friday or on the weekends (Monday thru Thursday only).

8.2 VERIFICATION

Verification that the plan is being correctly implemented requires review of documentation that control limits are being monitored and that corrective actions are being applied when CCP control limits are violated. Responsibility and accountability for this documentation is important. Independent verification of the HACCP plan is required.

Reassessment of the HACCP plan is an opportunity to simplify or expand as necessary. Effective HACCP teams typically reassess their plans once or twice a year.

Every hazard analysis and control plan should include process flow diagrams (Figures 6.4 and 6.5), a systematic hazard analysis and summary (Tables 6.5 and 6.6) the validation criteria that the hazard is controlled under operating conditions (Table 6.7) and a verification schedule (Table 6.8).

As an aid to reassessment of the hazard analysis and control plan, it is very helpful to organize around risk factors such as are given in risk classification schemes (e.g. Table 6.3). If the process has changed such that the critical risk criteria for any of the critical risk factors has resulted in risk classification into a lower category, then the hazard analysis and control plan can be adjusted accordingly. This is a way of scoring and documenting improvements that have resulted in less risk. A benefit for doing this is the revised hazard control plan may require less activity and therefore become easier to implement.

Conclusions

Legionellosis is the condition of being infected by *Legionella* bacteria. It causes serious disease that has resulted in many hundreds of thousands of deaths, has harmed millions of people and causes annual monetary losses in the billions.

Fundamentally, legionellosis is evidence of failure to properly maintain water systems. Legionellosis is the only cause of serious pneumonia that people get from contaminated building water systems. Most cases result from inadequate water treatment and poor water system management.

Legionellae are causative in severe pulmonary disease that can cause long-term or permanent damage. Debilitating effects to kidneys and in the nervous system are common in advanced disease. Therefore, curing the acute illness is not good enough. Prevention of the disease is necessary.

Victims contract the disease by aspirating (inhaling) water, aerosols or infective particles containing *Legionella* bacteria into their lungs. So far, there is no evidence for person-to-person transmission of the disease.

Legionellae are naturally intracellular parasites of aquatic protist hosts, mostly the amoebic protozoa. They are incidental parasites of mammalian phagocytes, mostly the alveolar macrophage (white blood cells).

Legionellosis disease is preventable. The biological hazard caused by *Legionella* bacteria in water systems can be eliminated, prevented or reduced to acceptable levels under operating conditions. This can be achieved by application of physical, chemical and biological hazard control technologies. Methods to safely and effectively apply these technologies are well-developed, widely available and cost-effective. Many thousands of legionellosis cases could be prevented each year if hazard control methods were used within the context of well-designed hazard analysis and control plans.

Legionellosis prevention requires a systematic approach, i.e., a process, to establish management accountability, assign responsibilities, describe the way water is processed in the facility, identify hazards, establish control limits for each critical point in the process, implement hazard control, monitor each control limit, document that the hazard has been controlled under operating conditions (validation) and independently confirm that the plan is being implemented (verification). These last two should be done to avoid commercial conflicts of interest that impede doing the right things to prevent legionellosis.

There are two conceptually equivalent hazard analysis and control processes that can be used equally well to organize all of these activities: They are the Water Safety Plan (WSP) or the Hazard Analysis and Critical Control Point (HACCP) system.

To prevent legionellosis we must:

- improve awareness and education
- implement water management plans
- mitigate against commercial conflicts of interest.

Bibliography

Abd, H., Johansson, T., Golovliov, I., Sandström, G. and Forsman, M. (2003) Survival and growth of *Francisella tularensis* in *Acanthamoeba castellanii*. *Appl. Environ. Microbiol.* **69**(1): 600–606.

Abu Kwaik, Y., Gao, L.Y., Stone, B.J., Venkataramna, C. and Harb, O.S. (1998) Invasion of protozoa by *Legionella pneumophila* and its role in bacterial ecology and pathogenesis. *Appl. Environ. Microbiol.* **64**(9): 3127–3133.

Addiss, D.G., Davis, J.P., La Venture, M., Wand, P.J., Hutchinson, M.A. and McKinney, R.M. (1989) Community-acquired Legionnaires' disease associated with a cooling tower: evidence for longer distance transport of *Legionella pneumophila*. *Am. J. Epidemiol.* **130**: 557–568.

Adeleke, A., Pruckler, J., Benson, R., Rowbotham, T., Halablab, M. and Fields, B.S. (1996) *Legionella*-like amoebal pathogens – phylogenetic status and possible role in respiratory disease. *Emer. Infect. Dis.* **2**: 225–229.

Alli, O.A.T., Zink, S., von Lackum, N.K. and Abu-Kwaik, Y. (2003) Comparative assessment of virulence traits in *Legionella* spp. *Microbiology* **149**: 631–641.

Angelillo, I.F., Viggiani, N.M., Greco, R.M. and Rito, D. (2001) HACCP and food hygiene in hospitals: knowledge, attitudes, and practices of food-services staff in Calabria, Italy. Collaborative Group. *Infect. Contr. Hosp. Epidemiol.* **22**(6): 363–369.

Aragon, V., Kurtz, S., McClain, M., Engleberg, N.C. and Cianciotto, N.P. (2002) Analysis of acid phosphatase and esterase/lipase mutants of *Legionella pneumophila*. In: *Legionella* (eds. Marre, R., *et al.*), Chapter 3, pp. 18–21. ASM Press, American Society for Microbiology, Washington, DC.

Armon, R., Starosvetsky, J., Arbel, T. and Green, M. (1997) Survival of *Legionella pneumophila* and *Salmonella typhimurium* in biofilm systems. *Water. Sci. Technol.* **35**(11–12): 293–300.

Arnow, P.M., Chou, T., Weil, D., Shapiro, E.N. and Kretzschmar, C. (1982) Nosocomial Legionnaires' disease caused by aerosolized tap water from respiratory devices. *J. Infect. Dis.* **146**: 460–467.

ASHRAE (American Society of Heating, Refrigeration and Air-Conditioning Engineers) (2000) Guideline 12–2000: Minimizing the Risk of Legionellosis Associated with Building Water Systems. 1791 Tullie Circle, N.E., Atlanta, GA 30329. *www.ashrae.org*

ASTM (American Society of Testing and Materials) D5952-96 (2002) Standard Guide for Inspecting Water Systems for *Legionellae* and Investigating Possible Outbreaks of Legionellosis (Legionnaires' Disease or Pontiac Fever). PO Box C700, Conshohocken, PA 19428-2959. *www.astm.org*

Atlas, R.M. (1999) *Legionella*: from environmental habitats to disease pathology, detection and control. *Environ. Microbiol.* **1**(4): 283–293.

Aurell, H., Etienne, J., Forey, F., Reyrolle, M., Girardo, P., Farge, P., Decludt, B., Campese, C., Vandenesch, F. and Jarraud, S. (2003) *Legionella pneumophia* serogroup 1 strain Paris: endemic distribution throughout France. *J. Clin. Microbiol.* **41**(7): 3320–3322.

Australian/New Zealand Standard (1995) Air-handling and water systems of buildings-microbial control. Part 1: Design, installation and commissioning. AS/NZS 3666.1:1995.

Australian/New Zealand Standard (1995) Air-handling and water systems of buildings-microbial control. Part 2: Operation and maintenance. AS/NZS 3666.2:1995.

Australian/New Zealand Standard (2000) Air-handling and water systems of buildings-microbial control. Part 3: Performance-based maintenance of cooling water systems. AS/NZS 3666.1:2000.

Australian/New Zealand Standard (2002) Air-handling and water systems of buildings-microbial control. Parts 1 and 2: Performance-based maintenance of cooling water systems. AS/NZS 3666.1:2002.

AWT (Association of Water Technologies) (2003) *Legionella* 2003: an update and statement by the association of water technologies (AWT). Association of Water Technologies, McLean, VA.

Bachman, M.A. and Swanson, M.S. (2004) Genetic evidence that *Legionella pneumophila* RpoS modulates expression of the transmission phenotype in both the exponential phase and the stationary phase. *Infect. Immun.* **72**(5): 2468–2476.

Barker, J., Brown, M.R.W., Collier, P.J., Farrell, I. and Gilbert, P. (1992) Relationship between *Legionella pneumophila* and *Acathamoeba polyphaga*: physiological status and susceptibility to chemical inactivation. *Appl. Environ. Microbiol.* **58**: 2420–2425.

Bartlett, C.L.R., Macrae, A.D., Macfarlane, J.T. (1986) Clinical aspects and diagnosis of *Legionella*. In: Legionella *Infections*, Chapter 3, p. 43, Fig. 3.1(a–d). Edward Arnold Publishers, London, UK. ISBN 0-7131-4506-4.

Benin, A.L., Benson, R.F., Arnold, K.E., Fiore, A.E., Cook, P.G., Williams, L.K., *et al.* (2002) An outbreak of travel-associated Legionnaires disease and Pontiac fever: The need for enhanced surveillance of travel-associated legionellosis in the United States. *J. Infect. Dis.* **185**: 237–243.

Benin, A.L., Benson, R.F. and Besser, R.E. (2002) Trends in Legionnaires' disease, 1980–1998: declining mortality and new patterns of diagnosis. *Clin. Infect. Dis.* 2002:**35** (1 November).

Benson, R.F., Tang, P.W. and Fields, B.S. (2000) Evaluation of the Binax and Biotest antigen kits for detection of Legionnaires' disease due to multiple serogroups and species of *Legionella*. *J. Clin. Microbiol.* **38**(7): 2763–2765.

Bentham, R.H., Pradham, M., Hakendorf, P. and Wilmot, P. (2002) Using geographical information systems for risk assessment and control of Legionnaires' disease associated with cooling towers. In: *Legionella* (eds. Marre, R., *et al.*), Chapter 63, pp. 318–320. ASM Press, American Society for Microbiology, Washington, DC.

Berk, S.G., Ting, R.S., Turner, G.W., and Ashburn, J. (1998) Production of respirable vesicles containing live *Legionella pneumophila* cells by two *Acanthamoeba* spp. *Appl. Environ. Microbiol.* **64**(1): 279–286.

Bergmans, D.C.J.J. and Bonten, M.J.M. (2004) Nosocomial pneumona. In: Hospital Epidemiology and Infection Control, pp. 311–339. Lippencott-Williams & Wilkins, Philadelphia, PA.

Birnbaum, H.G., Morley, M., Greenberg, P.E., Cifaldi, M. and Colice, G.L. (2001) Economic burden of pneumonia in an employed population. *Arch. Int. Med.* **161**: 2725–2731.

Birtles, R.J., Rowbotham, T.J., Michel, R., Pitcher, D.G., La Scola, B., Alexiou-Daniel, S., and Raoult, D. (2000) *Candidatus Odyssella thessalonicensis* gen. nov., sp. nov., and obligate intracellular parasite of *Acanthamoeba* species. *Int. J. Sys. Evol. Microbiol.* **50**: 63–72.

Bitar, D.M., Molmeret, M. and Abu, K.Y. (2004) Molecular and cell biology of *Legionella pneumophila. Int. J. Med. Microbiol.* **293**(7–8): 519–527.

Bollin, G.E., Plouffe, J.F., Para, M.F. and Hackman, B. (1985) Aerosols containing *Legionella pneumophila* generated by shower heads and hot water faucets. *Appl. Environ. Microbiol.* **50**: 1128–1131.

Borella, P., Montagna, M.T., Romano-Spica, V., Stampi, S., Stancanelli, G., Triassi, M., Neglia, R., Marchesi, I., Fantuzzi, G., Tato, D., Napoli, C., Quaranta, G., Laurenti, P., Leoni, E., De, L.G., Ossi, C., Moro, M. and Ribera, D.G. (2004) *Legionella* infection risk from domestic hot water. *Emer. Infect. Dis.* **10**(3): 457–464.

Bozue, J.A. and Johnson, W. (1996) Interaction of *Legionella pneumophila* with *Acanthamoeba castellanii*: uptake by coiling phagoytosis and inhibition of phagosome-lysosome fusion. *Infect. Immun.* **64**(2): 668–673.

Breiman, R.F. (1993) Modes of transmission in epidemic and nonepidemic *Legionella* infection: directions for further study, pp. 129–136. In: J.M. Barbaree, R.F. Breiman and A.P. Dufour (eds.), Legionella: *Current Status and Emerging Perspectives*. American Society for Microbiology, Washington, DC.

Brieland, J., McClain, M., Heath, L., Chrisp, C., Huffnagle, G., LeGendre, M., Hurley, M., Fantone, J. and Engleberg, C. (1996) Coinoculation with *Hartmannella vermiformis* enhances replicative *Legionella pneumophila* lung infection in a murine model of Legionnaires' disease. *Infect. Immun.* **64**(7): 2449–2456.

Brieland, J., McClain, M., LeGendre, M. and Engleberg, N.C. (1997b) Interpulmonary *Hartmannella vermiformis*: a potential niche for *Legionella pneumophila* replication in a murine model of legionellosis. *Infect. Immun.* **65**(1): 4892–4896.

Brieland, J.K., Fantone, J.C., Remick, D.G., LeGendre, M., McClain, M. and Engleberg, N.C. (1997a) The role of *Legionella pneumophila*-infected *Hartmannella vermiformis* as an infectious particle in a murine model of Legionnaires' disease. *Infect. Immun.* **65**(12): 5330–5333.

Brieland, J.K., Loebenberg, D., Menzel, F. and Hare, R.S. (2000) Efficacy of SCH27899 in an animal model of Legionnaires' disease using immunocompromised A/J mice. *Antimicrob. Agent. Chemother.* **44**(5): 1333–1336.

Broadbent, C. (1993) *Legionella* in cooling towers: practical research, design, treatment, and control guidelines, pp. 217–222. In: J.M. Barbaree, R.F. Breiman, and A.P. Dufour (eds.), Legionella: *Current Status and Emerging Perspectives*. American Society for Microbiology, Washington, DC.

Broadbent, C. (2002) Australian risk management approaches to control of legionella in cooling water systems, In: *Legionella* (eds., Marre, R., *et al.*), Chapter 371 pp. 371–375. ASM Press, American Society for Microbiology, Washington, DC.

Brown, M.R.W. and Barker, J. (1999) Unexplored reservoirs of pathogenic bacteria: protozoa and biofilms. *Trend. Microbiol.* **7**: 46–50.

Buchbinder, S., Trebesius, K. and Heesemann, J. (2002) Evaluation of detection of *Legionella* spp. in water samples by fluorescence in situ hybridization, PCR amplification and bacterial culture. *Int. J. Med. Microbiol.* **292**(3–4): 241–245.

Butler, J.C., Fields, B.S. and Brieman, R.F. (1997) Prevention and control of legionellosis. *Infect. Dis. Clin. Prac* **6**(7): 458–464.

Campos, C., Loret, J.F., Cooper, A.J. and Kelly, R.F. (2003) Disinfection of domestic water systems for *Legionella pneumophila. J. Water Suppl. Res.Tech.-AQUA.* (accepted for publication.)

CDC (Centers for Disease Control and Prevention), Healthcare Infection Control Practices Advisory Committee (2001) Draft Guideline for Environmental Infection Control in Healthcare Facilities. http://www.cdc.gov/ncidod/hip/enviro/guide.html.

CDC (Centers for Disease Control and Prevention), U.S. Department of Health and Human Services (2003) Summary of notifiable diseases – United States, 2001. *Morbid. Mortal. Weekly Report.* **50**(53) May 2, 2003.

CDC (Centers for Disease Control and Prevention) (2003) Guidelines for environmental infection control in health-care facilities: recommendations of CDC and the Healthcare Infection Control Practices Advisory Committee (HICPAC). MMWR; 52 (RR-10): 1–249. www.cdc.gov/ncidod/hip/enviro/guide.htm

CDC (Centers for Disease Control and Prevention) (1997) Final recommendations to minimize transmission of legionnaires' disease from whirlpool spas on cruise ships.

CDC (Centers for Disease Control and Prevention) (2000) Guidelines for preventing opportunistic infections among hematopoietic stem cell transplant recipients: recommendations of CDC, the Infectious Disease Society of America, and the American Society of Blood and Marrow Transplantation. *Morb Mortal Wkly Rep*; 49 (RR-10): 1–128. www.cdc.gov/mmwr/preview/mmwrhtml/rr4910a1.htm

CDC (Centers for Disease Control and Prevention) (2004) Guidelines for preventing health-care-associated pneumonia, 2003: recommendations of CDC and the Healthcare Infection Control Practices Advisory Committee (HICPAC). MMWR; 53 (RR-3): 1–179. www.cdc.gov/ncidod/hip/pneumonia/default.htm

CFR (US Code of Federal Regulations) (2002) Title 40 §141.52 Maximum contaminant level goals for microbiological contaminants. *www.epa.gov*

Chien, M., Morozova, I., Shi, S., *et al.* (2004) The genomic sequence of the accidental pathogen *Legionella pneumophila*. *Science* **305** (5692), 1966–1968.

CIBSE (Chartered Institution of Building Services Engineers) (2002) Minimizing the risk of Legionnaires' disease. TM13:2002. The Chartered Institution of Building Services Engineers, 222 Batham High Road., London SW 12 9BS.

Cianciatto, N., Eisenstein, B.I., Mody, C.H., Toews, G.B. and Engleberg, N.C. (1989) A *Legionella pneumophila* gene encoding a species-specific surface protein potentiates initiation of intracellular infection. *Infect. Immun.* **57**: 1255–1262.

Cianciotto, N.P., Kurtz, S., Krcmarik, K., Mody, C., Prasad, U., Robey, M., Salerno, J. and Viswanathan, V.K. (2002) Iron requirements of and acquisition of iron by *Legionella pneumophila,* In: *Legionella* (eds. Marre, R., *et al.*), Chapter 6, pp. 31–37. ASM Press, American Society for Microbiology, Washington, DC.

Cirillo, J.D., Cirillo, S.L., Yan, L., Bermudez, L.E., Falkow, S. and Tompkins, L.S. (1999) Intracellular growth in *Acanthamoeba castellanii* affects monocyte entry mechanisms and enhances virulence of *Legionella pneumophila*. *Infect. Immun.* **67**(9): 4427–4434.

Cloud, *et al.* (2000) Detection of *Legionella* species in respiratory specimens using PCR with sequencing confirmation. *J. Clin. Microbiol.* 1709–1712.

Cooper, A.J., Barnes, H.R. and Myers, E.R. (2004). Assessing risk of *Legionella*. *ASHRAE J.* **46**(4): 22–27.

Cordes, L.G., Faser, D.W., Skailiy, P., *et al.* (1980) Legionnaires' disease outbreak at an Atlanta, Georgia country club: evidence for spread from an evaporative condenser. *Am. J. Epidemiol.* **111**: 425–431.

Cordes, L.G., Weisenthal, A.M., Gorman, G.W. *et al.* (1981) Isolation of *Legionella pneumophila* from hospital shower heads. *Ann. Int. Med.* **94**: 195–197.

Costerton, J.W., Lewandowski, Z., Caldwell, D.E., Korber, D.R. and Lappin-Scott, H.M. (1995) Microbial biofilms. *Annu. Rev. Microbiol.* **49**: 711–745.

Craven, D.E. (2003) Progress in the battle against nosocomial Legionnaires' disease: shedding light on shades of gray. *Infect. Contr. Hosp. Epidemiol.* **24**(8): 560–562.

CTI (Cooling Technology Institute) (2000) Legionellosis. Guideline: best practices for control of *Legionella*. Cooling Technology Institute, Houston, TX 77273. www.cti.org

Cunha, B.A. (2003) Community-acquired pneumonia. *Infect. Med.* **20**(1): 27–30.

Dedicoat, M. and Venkatesan, P. (1999) The treatment of Legionnaires' disease. *J. Antimicr. Chemother.* **43**: 747–752.

den Boer, J.W., Yzerman, E.P.F., Schellekends, J., Lettinga, K.D., *et al.* (2002) A large outbreak of Legionnaires' disease at a flower show, the Netherlands, 1999. *Emer. Infect. Dis.* **8**(1): 37–43.

Dewettinck, T., Van Houtte, E., Geenens, D., Van Hege, K. and Verstraete, W. (2001) HACCP (Hazard Analysis and Critical Control Points) to guarantee safe water reuse and drinking water production – a case study. *Water Sci. Technol.* **43**(12): 31–38.

DHS (Department of Human Services, Public Health Division) (2001a) A guide to developing risk management plans. Government Department of Human Services, Melbourne, Victoria, Australia. www.legionella.vic.gov.au

DHS (Department of Human Services, Public Health Division) (2001b) Supplementary notes for hospitals. Public Health Group, Victorian Government Department of Human Services, Melbourne, Victoria. www.legionella.vic.gov.au

DHS (Department of Human Services, Public Health Division) (2002) Code of practice for water treatment service providers (cooling tower systems). Public Health Group, Victorian Government Department of Human Services, Melbourne, Victoria, Australia. *www.legionella.vic.gov.au*

DIN (Deutsches Institut fur Normung) (1997) DIN 19643-1 and 38402-14.

Dondero, T.J., Rentdorff, R.C., Mallison, G.F., Weeks, R.M., Levy, J.S., Wong, E.W. and Schaffner, W. (1980) An outbreak of Legionnaires' disease associated with a contaminated air-conditioning cooling tower. *New Engl. J. Med.* **302**: 365–370.

Donlan, R.M. and Costerton, J.W. (2002) Biofilms: survival mechanisms of clinically relevant microorganisms. *Clin. Microbiol. Rev.* **15**(2): 167–193.

Donlan, R.M. (2002) Biofilms: microbial life on surfaces. *Emer. Infect. Dis.* **8**: 881–890.

Donlan, R.M., Forster, T., Murga, R. Brown, E., Lucas, C., Carpenter, J and Fields, B. (2005) *Legionella pneumophila* associated with the protozoan *Hartmannella vermiformis* in a model multi-species biofilm has reduced susceptibility to disinfectants. *Biofouling* **21**(1): 1–7.

Dunbar, L.M., Khashab, M.M., Kahn, J.B., Zadeikis, N., Xiang, J.X. and Tennenberg, A.M. (2004) Efficacy of 750-mg, 5-day levofloxacin in the treatment of community-acquired pneumonia caused by atypical pathogens. *Curr. Med. Res. Opin.* **20**(4): 555–563.

DVGW (Deutscher Verein des Gas- und Wasserfaches e.V) (1996) Hot water systems and pipes; technical improvements for the reduction of the growth of *Legionella*: renovation and operation. Manual W 551 and W 552.

Eckmanns, T., Schwab, F., Posselt, H., Gastmeier, P. and Rüden, H. (2002) UV light for elimination of Legionellae In: *Legionella* (eds. Marre, R. *et al.*), Chapter 81, pp. 402–405. ASM Press, American Society for Microbiology, Washington, DC.

Edelstein, P.H. (2002) Chemotherapy of Legionnaires' disease with macrolide or quinolone antimicrobial agents. In: *Legionella* (eds. Marre, R. *et al.*), Chapter 32, pp. 183–188. ASM Press, American Society for Microbiology, Washington, DC.

EPA (US Environmental Protection Agency) (1999) *Alternative Disinfectants and Oxidants Guidance Manual.* EPA 815-R-99-014, Office of Water (4607), April 1999. www.epa.gov

EPA (US Environmental Protection Agency) (2001) *Legionella: Drinking Water Health Advisory.* EPA 822-B-01-005, Office of Science and Technology, March 2001. www.epa.gov

Essig, A., Heinemann, M., Simnacher, U. and Marre, R. (1997) Infection of *Acanthamoeba castellanii* by *Chlamydia pneumoniae. Appl. Environ. Microbiol.* **63**: 1396–1399.

Eurovent 9/5 (2000) Guideline for the prevention of bacteriological contamination, including *Legionella pneumophila*, in cooling towers and evaporative condensers.

Fabbi, M., Pastoris, M.C., Scanziani, E., Magnino, S. and Di, M.L. (1998) Epidemiological and environmental investigations of *Legionella pneumophila* infection in cattle and case report of fatal pneumonia in a calf. *J. Clin. Microbiol.* **36**(7), 1942–1947.

Faulkner, G. and Garduño, A. (2002) Ultrastructural analysis of differentiation in *Legionella pneumophila*. *J. Bact.* **184**(24): 7025–7041.

Feeley, J.C., Gorman, R.E., Weaver, R.E., Mackel, D.C. and Smith, H.W. (1978) Primary isolation media for Legionnaires' disease bacterium. *J. Clin. Microbiol.* **8**: 320–325.

Fields, B.S. (1996) The molecular ecology of legionellae. *Trend. Microbiol.* **4**: 286–290.

Fields, B.S. (1993) *Legionella* and protozoa: interaction of a pathogen and its natural host, In: Legionella: *Current Status and Emerging Perspectives.* (eds. Barbaree, J.M., Breiman, R.F. and Dufour, A.P.), pp. 129–136. American Society for Microbiology, Washington, DC.

Fields, B.S., Benson, R.F. and Besser, R.E. (2002) *Legionella* and Legionnaires' disease: 25 Years of Investigation. *Clin. Microbiol. Rev.* **15**(3): 506–526.

Fields, B.S., Sanden, G.N., Barbaree, J.M. *et al.* (1989) Intracellular multiplication of *Legionella pneumophila* in amoebae isolated from hospital hot water tanks. *Curr. Microbiol.* **18**: 131–137.

Flieger, A., Gong, S., Faigle, M., Stevanovic, S., Northoff, H., Cianciotto, N.P. and Neumeister, B. (2002) *Legionella pneumopnila* secretes different phosholipases A In: *Legionella* (eds. Marre, R. *et al.*), Chapter 4, pp. 22–26. ASM Press, American Society for Microbiology, Washington, DC.

Forsbach-Birk, V., McNealy, T., Shi, C., Lynch, D. and Marre, R. (2004) Reduced expression of the global regulator protein CsrA in *Legionella pneumophila* affects virulence-associated regulators and growth in *Acanthamoeba castellanii. Int. J. Med. Microbiol.* **294**(1): 15–25.

France (2005a) Decrees, minutes, circulations. Minutes from the December 13, 2004 assembly of refrigeration installation by water dispersion in air supply in submission to authorization title heading of number 2921. Authorized February, 2005 *Ministère de l'écologie et du développement durable,* Editions Copédit/ICS'eau – 2005.

France (2005b) Guide méthodologique pour la réalisation d'une analyse de risque de prolifération de légionelles dans les installations de refroidissement par dispersion d'eau dans un flux d'air. *Ministère de l'écologie et du développement durable.* République Française.*www1.environnement.gouv.fr/IMG/pdf/GUIDEWEB_analyse_de_risque_de_proliferation_des_legionelles.pdf*

France (2005c) Dossier sur la légionellose. *Ministère de l'écologie et du développement durable.* République Française. *www.ecologie.gouv.fr/rubrique.php3?id_rubrique = 936).#*

France (2001) Guide des bonnes pratiques: *Legionella* et tours aéroréfrigérantes. *Ministére de L'Emploi et de la Solidartité, Ministére de L'Economie des Finances et de L'Industrie, Ministére de L'Amenagement du Territoire et de L'Environment.*

Franceschini, N., Boschi, L., Pollini, S., Herman, R., Perilli, M., Galleni, M., Frere, J.M., Amicosante, G. and Rossolini, G.M. (2001) Characterization of OXA-29 from *Legionella* (*Fluoribacter*) gormanii: molecular class D beta-lactamase with unusual properties. *Antimicrob. Agent Chemother.* **45**(12): 3509–3516.

Franzin, L., Cabodi, D. and Fantino, C. (2002) Evaluation of the efficacy of ultraviolet irradiation for disinfection of hospital water contaminated by *Legionella. J. Hosp. Infect.* **51**(4): 269–274.

Fraser, D.W. (1980) Legionellosis: evidence of airborne transmission. *Ann. N.Y. Acad. Sci.* **353**: 61–66.

Fraser, D.W. (1984) Sources of legionellosis, pp. 277–280. In: Thornsberry, C., Balows, A., Feeley, J.C. and Jakubowski, W (eds.), *Legionella*: *Proceedings of the Second International Symposium,* American Society for Microbiology, Washington, DC.

Fraser, D.W., Tsai, T.R., Orenstein, W., Parkin, W.E., Beecham, H.J., Sharrar, R.G., *et al.* (1977) Legionnaires' disease: description of an epidemic of pneumonia. *New Engl. J. Med.* **297**: 1189–1197.

Fritsche, T.R., Horn, M., Seyedirashti, S., Gautom, R.K., Schleifer, K-H. and Wagner, M. (1999) *In situ* detection of novel bacterial endosymbionts of *Acanthamoeba* spp. phylogenetically related to members of the order *Rickettsiales. Appl. Environ. Microbiol.* **65**(1): 206–212.

Fung, A.S., Leikis, M.J. and McMahon, L.P. (2004) Acute renal failure associated with *Legionella pneumonia* and acute cholecystitis. *Nephrology (Carlton.)* **9**(2): 105–108.

Garbe, P.L., Davis, B.J., Weisfield, J.S., Markowitz, L., Miner, P., Garrity, F., Barbaree, J.M. and Reingold. A.L. (1985) Nosocomial Legionnaires' disease: epidemiologic demonstration of cooling towers as a source. *JAMA* **254**: 521–524.

García-Fulgueiras, A., Navarro, C., Fenoll, D., García, J., *et al.* (2003) Legionnaires' disease outbreak in Murcia, Spain. *Emer. Infect. Dis.* **9**(8) August 2003. Available from: URL: http://www.cdc.gov/ncidod/EID/vol0no8/03-0337.htm

Garduno, R.A., Faulkner, G., Trevors M.A., Vats, N. and Hoffman, P.S. (1998) Immunolocalization of Hsp60 in *Legionella pneumophila*. *J. Bacteriol.* **180**(3): 505–513.

Goetz, A.M., Stout, J.E., Jacobs, S.L., Fisher, M.A., Ponzer, R.E., Drenning, S., and Yu, V.L. (1998) Nosocomial legionnaires' disease discovered in community hospitals following cultures of the water system: seek and ye shall find. *Am. J. Infect. Contr.* **26**(1): 8–11.

Greig, J.E., Carnie, J.A., Tallis, G.F., Ryan, N.J., Tan, A.G., Gordon, I.R., Zwolak, B., Leydon, J.A., Guest, C.S. and Hart, W.G. (2004) An outbreak of Legionnaires' disease at the Melbourne Aquarium, April 2000: investigation and case–control studies. *Med. J. Australia.* **180**(11): 566–572.

Greub, G. and Raoult, D. (2002) Crescent bodies of *Parachlamydia acanthamoeba* and its life cycle within *Acanthamoeba polyphaga*: an electron micrograph study. *Appl. Environ. Microbiol.* **68**(6): 3076–3084.

Grosserode, M., *et al.* (1993) Continuous huperchlorination for control of nosocomial *Legionella pneumophila* pneumonia: a ten-year follow-up of efficacy, environmental effects, and dosts. In: Legionella: *Current Status and Emerging Perspectives* (eds., Barbarre, J.M., Breiman, R.F., Dufour, A.P.), pp. 226–229. ASM Press, American Society for Microbiology, Washington, DC.

Guest, J.F. and Morris, A. (1997) Community-acquired pneumonia: the annual cost to the National Health Service in the UK. *Eur. Respir. J.* **10**(7): 1530–1534.

Gutell, R.R., Larsen, N. and Woese, C.R. (1994) Lessons from an evolving rRNA: 16S and 23S rRNA structures from a comparative perspective. *Microbiol. Rev.* **58**(1): 10–26.

Habicht, W. and Muller, H.E. (1988) Occurrence and parameters of frequency of *Legionella* in warm water systems of hospitals and hotels in Lower Saxony. *Pravent. Medzin* (written in English) **188**(1): 79–88.

Hall, K.K., Giannetta, E.T., Getchell-White, S.I., *et al.* (2003) Ultraviolet light disinfection of hospital water for preventing nosocomial *Legionella* infection: a 13-year follow-up. *Infect. Contr. Hosp. Epidemiol.* **24**(8): 580–583.

Hammer, B.K. and Swanson, M.S. (1999) Co-ordination of *Legionella pneumophila* virulence with entry into stationary phase by ppGpp. *Mol. Microbiol.* **33**: 721–731.

Hanrahan, J.P., Morse, D.L. and Scharf, V.B. (1987) A community hospital outbreak of legionellosis: transmission by potable hot water. *Am. J. Epidemiol.* **125**: 639–649.

Hawn, T.R., Verbon, A., Lettinga, K.D., Zhao, L.P., Li, S.S., Laws, R.J., Skerrett, S.J., Beutler, B., Schroeder, L., Nachman, A., Ozinsky, A., Smith, K.D. and Aderem, A. (2003) A common dominant TLR5 stop codon polymorphism abolishes flagellin signaling and is associated with susceptibility to legionnaires' disease. *J. Exp. Med.* **198**(10): 1563–1572.

Hayden, R.T., Uhl, J.R., Qian, X., Hopkins, M.K., Aubry, M.C., Limper, A.H., Lloyd, R.V. and Cockerill, F.R. (2001) Direct detection of *Legionella* species from bronchoalveolar lavage and open lung biopsy specimens: comparison of LightCycler PCR, in situ hybridization, direct fluorescence antigen detection, and culture. *J. Clin. Microbiol.* **39**(7): 2618–2626.

Heath, L., Chrisp, C., Huffnagle, G., LeGendre, M., Osawa, Y., Hurley, M., Engleberg, C., Fantone, J. and Brieland, J. (1996) Effector mechanisms responsible for gamma interferon-mediated host resistance to *Legionella pneumophila* lung infection: the role of endogenous nitric oxide differs in susceptible and resistant murine hosts. *Infect. Immun.* **64**(12): 5151–5160.

Heffelfinger, J.D., Kool, J.L., Fridkin, S., *et al.* (2003) Risk of hospital-acquired Legionnaires' disease in cities using monochloramine versus other water disinfectants. *Infect. Contr. Hosp. Epidemiol.* **24**(8): 569–574.

Helbig, J.H., Lück, C., Jacobs, E. and Witt, M. (2002) Localization of *Legionella pneumophila* Mip protein inside phagosomes of *Acanthamoeba castellanii* In: *Legionella* (eds. Marre, R., *et al.*), Chapter 9, pp. 49–51 ASM Press, American Society for Microbiology, Washington, DC.

Heng, B.H., Goh, K.T., Ng, D.L.K. and Ling, A.E. (1997) Surveillance of Legionellosis and *Legionella* bacteria in the built environment in Singapore. *Ann. Acad. Med. Singapore.* **26**: 225–565.

Herpers, B.L., de Jongh, B.M., van der, Z.K. and van Hannen, E.J. (2003) Real-time PCR assay targets the 23S-5S spacer for direct detection and differentiation of *Legionella* spp. and Legionella pneumophila. *J. Clin. Microbiol.* **41**(10): 4815–4816.

Heuner, K., Steinert, M., Dietrich, C., Fischer, G., Köhler, R. and Hacker, J. (2002) Function and expression of *Legionella pneumophila* surace factors. In: *Legionella* (eds. Marre, R. *et al.*), Chapter 8, pp. 43–48 ASM Press, American Society for Microbiology, Washington, DC.

Hlady, W.G., Mullen, R.C., Mintz, C.S., Shelton, B.G., Hopkins, R.S. and Daikos, G.L. (1993) Outbreak of Legionnaires' disease linked to a decorative fountain by molecular epidemiology. *Am. J. Epidemiol.* **138**: 555–562.

Hong Kong (2000) Prevention of Legionnaires' disease. The Government of Hong Kong. www.ab.ust.hk/sepo/pdf/legio.pdf

Horman, A., Rimhanen-Finne, R., Maunula, L., von Bonsdorff, C.H., Rapala, J., Lahti, K. and Hanninen, M.L. (2004) Evaluation of the purification capacity of nine portable, small-scale water purification devices. *Water Sci. Technol.* **50**(1): 179–183.

Horn, M., Wagner, M., Müller, K.-D., Schmid, E.N., Fritsche, T.R., Schleifer, K.-H. and Michel, R. (2000) *Neochlamydia hartmannellae* gen. nov., sp. nov. (*Parachlamydiaceae*), an endoparasite of the amoeba *Hartmannella vermiformis*. *Microbiology* **146**: 1231–1239.

Howard, G. (2003) Water safety plans for small systems: a model for applying HACCP concepts for cost-effective monitoring in developing countries. *Water Sci. Technol.* **47**(3) 215–220

HSC (Health and Safety Commission and Executive) (2000) Legionnaires' disease. The Control of Legionella Bacteria in Water Systems. Approved Code of Practice and Guidance, L8. *HSE Books*. PO Box 1999, Sudbury, UK. *www.hse.gov.uk*

Huang, B., Heron, B.A., Gray, B.R., Eglezos, S., Bates, J.R. and Savill, J. (2004) A predominant and virulent Legionella pneumophila serogroup 1 strain detected in isolates from patients and water in Queensland, Australia, by an amplified fragment length polymorphism protocol and virulence gene-based PCR assays. *J. Clin. Microbiol.* **42**(9): 4164–4168.

Inglis, T.J.J., Rigby, P., Robertson, T.A., Dutton, N.S., Henderson, M. and Chang, B.J. (2000) Interaction between *Burkholderia pseudomallei* and *Acanthamoeba* species results in coiling phagocytosis, endamebic bacterial survival, and escape. *Infect. Immun.* **68**(3): 1681–1686.

Inoue, H., Kawano, G., Nagasawa, H. and Sakuda, S. (2002) Isolation of elemental sulfur as a self-growth-inhibiting substance produced by *Legionella pneumophila*. *Appl. Environ. Microbiol.* **68**(10): 4809–4811.

ISO (1998) Water quality-Detection and enumeration of *Legionella*. ISO 11731. *International Organization for Standardization*, Geneva, Switzerland.

Jagals, C. and Jagals, P. (2004) Application of HACCP principles as a management tool for monitoring and controlling microbiological hazards in water treatment facilities. *Water Sci. Technol.* **50**(1): 69–76.

JCAHO (Joint Commission on Accreditation of Healthcare Organizations) (2001) Utility Systems Management. Std EC.1.7. *www.jcaho.org/Onlineform/OnLineForm.asp*

Jernigan, D.B., Hofmann, J., Cetron, M.S., *et al.* (1996) Outbreak of Legionnaires' disease among cruise ship passengers exposed to a contaminated whirlpool spa. *Lancet* **347**: 494–499.

Joly, J.R. (1993) Prevention and control of legionellosis. In: *Legionella: Current Status and Emerging Perspectives*. (eds.) Barbaree, J.M., *et al.* (eds.). American Society for Microbiology, Washington, DC.

Joseph, C. (2002) Surveillance of Legionnaires' Disease in Europe. In: *Legionella* (eds., Marre, R., *et al.*), Chapter 62, pp. 311–317 ASM Press, American Society for Microbiology, Washington, DC.

Joseph, C.A. (2004) Legionnaires' disease in Europe 2000–2002. *Epidemiol. Infect.* **132**: 417–424.

Joshi, A. and Swanson, M.S. (1999) Comparative analysis of *Legionella pneumophila* and *Legionella micadadei* virulence traits. *Infect. Immun.* **67**(8): 4134–4142.

Junge-Mathys, E. and Mathys, W. (1994) Die Legionellose – ein Beispiel für umweltbedingte Infektionen [Legionellosis – an example of environmentally caused infections]. *Intensiv*, 2: 29–33.

Kahane, S., Dvoskin, B., Mathias, M. and Friedman, M. (2001) Infection of *Acanthamoeba polyphaga* with *Simkania negevensis* and *S. negevensis* survival with amoebal cysts. *Appl. Environ. Microbiol.* **67**(10): 4789–4795.

Keller, D.W., Hajjeh, R., DeMaria, A., *et al.* (1996) Community outbreak of Legionnaires' disease: an investigation confirming the potential for cooling towers to transmit *Legionella* species. *Clin. Infect. Dis.* **22**: 257–261.

Khelef, N., Shuman, H.A. and Maxfield, F.R. (2001) Phagocytosis of wild-type *Legionella pneumophila* occurs through a wortmannin-insensitive pathway. *Infect. Immun.* **69**(8): 5157–5161.

Kilvington, S. and Price, J. (1990) Survival of *L. pneumophila* within cysts of *Acanthamoeba polyphaga* following chlorine exposure. *J. Appl. Bacteriol.* **68**: 519–525.

Kim, B.R., Anderson, J.E., Mueller, S.A., Gaines, W.A. and Kendall, A.M. (2002) Literature review-efficacy of various disinfectants against *Legionella* in water systems. *Water Res.* **36**: 4433–4444.

King, C.H., Shotts Jr., E.B., Wooley, R.E. and Porter, K.G. (1988) Survival of coliforms and bacterial pathogens within protozoa during chlorination. *Appl. Environ. Microbiol.* **54**: 3023–3033.

Kistemann, T., Herbst, S., Dangendorf, F. and Exner, M. (2001) GIS-based analysis of drinking-water supply structures: a module for microbial risk assessment. *Int. J. Hyg. Environ. Health.* **203**(4): 301–310.

Klahre, J. and Flemming, H-C. (2000) Monitoring of biofouling in papermill process waters. *Water. Res.* **34**: 3657–3665.

Koide, M., Saito, A., Kusano, N. and Higa, F. (1993) Detection of *Legionella* spp. in cooling tower water by the polymerase chain reaction method. *Appl. Environ. Microbiol.* **59**: 1943–1946.

Kool, J.L., Carpenter, J.C. and Fields B.S. (1999) Effect of monochloramine disinfection of municipal drinking water on risk of nosocomial Legionnaires' disease. *Lancet* **353**(9149): 272–277.

Kroca, M., Johansson, A., Sjostedt, A. and Tarnvik, A. (2001) V gamma 9V delta 2 T cells in human legionellosis. *Clin. Diagn. Lab. Immunol.* **8**(5): 949–954.

Kusnetsov, J.M., Martikainen, P.J., Pertti, J., *et al.* (1997) Physical, chemical and microbiological water characteristics associated with the occurrence of *Legionella* in cooling tower systems. *Water Res.* **27**(1): 85–90.

Kusnetsov, J., Torvinen, E., Perola, O., Nousiainen, T. and Katila, M.L. (2003) Colonization of hospital water systems by legionellae, mycobacteria and other heterotrophic bacteria potentially hazardous to risk group patients. *APMIS* **111**(5): 546–556.

La Scola, B., Barrassi, L. and Raoult, D. (2000) Isolation of new fastidious α-proteobacteria and *Afipia felis* from hospital water supplies by direct plating and amoebal co-culture procedures. *FEMS Microbiol. Ecol.* **34**: 129–137.

La, S.B., Boyadjiev, I., Greub, G., Khamis, A., Martin, C. and Raoult, D. (2003) Amoeba-resisting bacteria and ventilator-associated pneumonia. *Emer. Infect. Dis.* **9**(7): 815–821.

Lave, J.R., Lin, C.J., Fine, M.J. and Hughes-Cromwick, P. (1999) The cost of treating patients with community-acquired pneumonia. *Semin. Respir. Crit. Care Med.* **20**(3): 189–197.

Lee, J.V., Surman, S.B., Kirby, A. and Seddon, F. (2002) Eleven years of experience with novel strategies for *Legionella* control in a large teaching hospital. In: *Legionella*, (eds., Marre, R. *et al.*), Chapter 80, pp. 398–401. ASM Press, American Society for Microbiology, Washington, DC.

Lee, V.T. and Schneewind, O. (2001) Protein secretion and the pathogenesis of bacterial infections. *Gene Develop.* 15: 1725–1752.

Lim, W.S. Slack, R. Goodwin, A. Robinson, J. Lee, J.V. Joseph, C. and Neal, K. (2003) Community-acquired Legionnaires' disease in Nottingham – too many cases? Epidemiol. Infect. **131**(3): 1097–1103.

Lin, Y.E., Vidic, R.D., Stout, J.E. and Yu, V.L. (1998) *Legionella* in water distribution systems. *J. Am. Water Work. Assoc.* **90**(9): 112–121.

Lin, Y.E., Vidic, R.D., Stout, J.E. and Yu, V.L. (2002) Negative effect of high pH on biocidal efficacy of copper and silver ions in controlling *Legionella pneumophila*. *Appl. Environ. Microbiol.* **68**(6): 2711–2715.

Loret, J.-F., Robert, S., Thomas, Cooper, A.J., McCoy., W.F. and Lévi, V. (2005) Comparison of disinfectants for biofilm, protozoa and *Legionella* control. *J. Water Health* (accepted for publication).

Lück, P.C., *et al.* (1993) Prevalence of *Legionella* species, serogroups and monoclonal subgroups in hot water systems in south-eastern Germany. *Zeitblatt für Hygiene Umweltmedizin*, **193**: 450–460.

Ly, T.M. and Müller, H.E. (1990) Ingested *Listeria monocytogenes* survive and multiply in protozoa. *J. Med. Microbiol.* **33**: 129–137.

Lye, D., Fout, G. S., Crout S.R., Danielson, R., Thio, C.L. and Paszko-Kolva, C.M. (1997) Survey of ground, surface, and potable waters for the presence of *Legionella* species by enviroampR PCR *Legionella* kit, culture, and immunofluorescent staining. *Water. Res.* **31**(2): 287–293.

Mahoney, F.J., Hoge, C.W., Farley, T.A., *et al.* (1992) Community-wide outbreak of Legionnaires' disease associated with a grocery store mist machine. *J. Infect. Dis.* **165**: 736–739.

Mangione, E.J., Remis, R.S., Tait, K.A., McGee, W.B., Gorman, G.W., Wentworth, B.B., Baron, P.A., Hightower, A.W., Barbaree, J.M. and Broome, C.V. (1982) An outbreak of Pontiac fever related to whirlpool use, Michigan. *JAMA* **253**: 535–539.

Marciano-Cabral, F. and Cabral, G. (2003) *Acanthamoeba* spp. as agents of disease in humans. *Clin. Microbiol. Rev.* **16**(2): 273–307.

Marre, R., Medeiros, A.A. and Pasculle, A.W. (1982) Characterization of the β-lactamases of six species of *Legionella*. *J. Bacteriol.* **151**(1): 212–216.

Marrie, T., Green, P., Burbridge, S., Bezanson, G., Neale, S., Hoffman, P.S., *et al.* (1994) *Legionellaceae* in the potable water of Nova Scotia hospitals and Halifax residences. *Epidemiol. Infect.* **112**: 143–150.

Marston, B.J., Lipman, H.B. and Breiman, R.F. (1994) Surveillance for Legionnaires' disease. *Arch. Int. Med.* **154**: 2417–2422.

Marston, B.J., Plouffe, J.F., File, T.M., Hackman, B.A., Salstrom, S.J., Lipman, H.B., Kolczak, M.S. and Breiman, R.F. (1997) Incidence of community-acquired pneumonia requiring hospitalization. Results of a population-based active surveillance study in Ophon. The commuity-based pneumonia incidence study group. *Arch. Int. Med.* **157**(15): 1709–1718.

Maryland, Department of Health and Mental Hygiene (2000) Report of the Maryland Scientific Working Group to Study *Legionella* in Water Systems in Healthcare Institutions, pp. 1–27. www.dhmh.state.md.us/html/legionella.htm

Mastro, T.D., Fields, B.S., Breiman, R.F., Campbell, J., Plikaytis, B.D. and Spika, J.S. (1991) Nosocomial Legionnaires' disease and use of medication nebulizers. *J. Infect. Dis.* **163**: 667–670.

Maurin, M. and Raoult, D. (1999) Q fever. *Clin. Microbiol. Rev.* **12**(4): 518–553.

McAnulty, J., Habib, M. and Delpech, V. (2001) Epireview: Legionairres' disease, NSW, 1991–2000. *NSW Public Health Bulletin* **12**(10): 282–285 www.health.nsw.gov.au

McCoy, W.F. (2004). *Legionella*. In: Microbial Waterborne Pathogens (eds. Cloete, E., *et al.*), Chapter 5 IWA Publishing and The World Health Organization. ISBN: 1843390558

McCoy, W.F. (2003a) Microbial fouling control for cooling water. In: *Business Briefing: CPI Technology*, pp. 65–67. World Markets Research Centre, Ltd, WMRC plc, London, UK. *www.wmrc.com*

McCoy, W.F. (2003b) Microbial-fouling control for industrial systems. In: *Biofilms in Medicine, Industry and Environmental Biotechnology: Characteristics, Analysis and Control* (eds. Lens, P., *et al.*) IWA Publishing, London.

McDade, J.E., Sheppard, C.C., Faser, D.W., *et al.* (1977) Legionnaires' disease: isolation of a bacterium and demonstration of its role in other respiratory diseases. *New Engl. J. Med.* **287**: 1197–1203.

Mealey (2001) $9.6 million award affirmed in New York Federal Court for Legionnaire's (sic) disease. *Mealey's Daubert Report*. 11 (2001), **5**(11). Published by LexisNexis, a division of Reed Elsevier Inc.

Michel, R., Muller, K.D. and Schmid, E.N. (1995) *Ehrlichia*-like organisms (KSL1) observed as obligate intracellular parasites of *Saccamoeba* species. *Endocytobiosis Cell Res.* **11**: 69–80.

Miller, R.D. and Kenepp, K.A. (1993) Risk assessments for *Legionnaires'* disease based on routine surveillance of cooling towers for Legionellae. In: Legionella: *Current Status and Emerging Perspectives*. (eds., Barbaree, J.M., *et al.*), American Society for Microbiology. Washington, D.C.

Miller, R.D. (2002) Prevalence of *Legionella* in whirlpool spas: Correlation with total bacterial numbers. In: Legionella: *Current Status and Emerging Perspectives.* (eds., Barbaree, J.M., Breiman, R.F. and Dufour, A.P.) American Society for Microbiology, Washington, DC.

Molmeret, M., Alli, O.A.T., Zink, S., Flieger, A., Cianciootto, N.P. and Kwaik, Y.A. (2002) IcmT is essential for pore formation-mediated egress *of Legionella pneumopnila* from mammalian and protozoan cells. *Infect. Immun.* **70**(1): 69–78.

Molofsky, A.B. and Swanson, M.S. (2004) Differentiate to thrive: lessons from the *Legionella pneumophila* life cycle. *Mol. Microbiol.* **53**(1): 29–40.

Muder, R.R. and Yu, V.L. (2002) Infection due to *Legionella* species other than L. pneumophila. *Clin. Infect. Dis.* **35**(8): 990–998.

Muder, R.R., Stout, J.E. and Yu, V.L. (2000) Nosocomial *Legionella micdadei* infection in transplant patients: fortune favors the prepared mind. *Am. J. Med.* **108**(4): 346–348.

Muder, R.R., Yu, V.L. and Woo, A.H. (1986) Mode of transmission of *Legionella pneumophila*. A critical review. *Arch. Int. Med.* **146**(8): 1607–1612.

Mulazimoglu, L. and Yu, V.L. (2001). Can Legionnaires' disease be diagnosed by clinical criteria: a critical review. *Chest* **120**(4): 1049–1053.

Murga, R., Forster, T.S., Brown, E., Pruckler, J.M., Fields, B.S. and Donlan, R.M. (2001) Role of biofilms in the survival of *Legionella pneumophila* in a model potable-water system. *Microbiology* **147**: 3121–3126.

Nadebaum, P., Chapman, M., Ortisi, S. and Baker, A. (2003) Application of quality management systems for drinking water quality. *Water Suppl.* **3**(1–2): 359–364.

Nagal, H. and Roy, C.R. (2003) Show me the substrates: modulation of host cell function by type IV secretion systems. *Cell. Microbiol.* **5**(6): 373–383.

Nahapetian, K., Challemel, O., Beurtrin, D., Burbrou, S., Gournon, P. and Squinazi, F. (1991) The intracellular multiplication of *Legionella pneumophila* in protozoa from hospital plumbing systems. *Res. Microbiol.* **142**: 677–685.

Neumeister, B., Schoniger, S., Faigle, M., Eichner, M. and Dietz, K. (1997) Multiplication of different Legionella species in Mono Mac 6 cells and in *Acanthamoeba castellanii*. *Appl. Environ. Microbiol.* **63**(4): 1219–1224.

Neuner, A.R. (2002) Health system triumphs over *Legionella*. *AFE Facil. Eng. J.* Nov/Dec 2002. *www.afe.org*

Newsome, A.L., Scott, T.M., Benson, R.F. and Fields, B.S. (1998) Isolation of an amoeba naturally harboring a distinctive *Legionella* species. *Appl. Environ. Microbiol.* **64**(5): 1688–1693.

Olsen, G.J., Woese, C.R. and Overbeek, R. (1994) The winds of (evolutionary) change: breathing new life into microbiology. *J. Bacteriol.* **176**(1): 1–6.

OSHA (Occupational, Safety and Health Administration) (1998) Technical Manual, Section II, Chapter 7, Legionnaires' disease. *www.osha-slc.gov/TechMan_data/II_7.html*

PA, Allegheny County Health Department. (1997) Approaches to the prevention and control of *Legionella* infection in Allegheny County Health Care Facilities, 2nd ed. Allegheny County Health Department, Pittsburgh, PA.

Palmer, C.J., Bonilla, G.F., Roll, B., Paszko-Kolva, C., Sangermano, L.R. and Fujioka, R.S. (1995) Detection of *Legionella* species in reclaimed water and air with the EnviroAmp Legionella PCR kit and direct fluorescent antibody staining. *Appl. Environ. Microbiol.* **61**: 407–412.

Palmer, C.J., Tsai, Y.L., Paszko-Kolva, C., Mayer, C. and Sangermano, L.R. (1993) Detection of *Legionella* species in sewage and ocean water by polymerase chain reaction, direct fluorescent-antibody, and plate culture methods. *Appl. Environ. Microbiol.* **59**: 3618–3624.

Percival, S.T., Walker, J.T. and Hunter, P.R (2000) Risk assessment, Chapter 4, page 41–48. In: *Microbiological Aspects of Biofilms and Drinking Water.* CRC Press, London.

Riffard, S., Douglass, S., Brooks, T., Springthorpe, S., Filion, L.G. and Sattar, S.A. (2001) Occurrence of *Legionella* in groundwater: an ecological study. *Water Sci. Tech.* **43**(12): 99–102.

Riffard, S., Springthorpe, S., Filion, L., Sutter, S., Brooks, T., Osicki, R., Lee, M. and Abrial, D. (2005) Occurrence of *Legionella* in groundwater. ISBN: 1843398818. IWA Publishing.

Robey, M., O'Connell, W. and Cianciotto, N.P. (2001) Identification of *Legionella pneumophila* rcp, a pagP-like gene that confers resistance to cationic antimicrobial peptides and promotes intracellular infection. *Infect. Immun.* **69**(7): 4276–4286.

Rohr, U., Weber, S., Michel, R., Selenka, F. and Wilhelm, M. (1998) Comparison of free-living amoebae in hot water systems of hospitals with isolates from moist sanitary areas by identifying genera and determining temperature tolerance. *Appl. Environ. Microbol.* **64**(5): 1822–1824.

Rose, E. (2005) UK Debates proposed law: corporate manslaughter. ASHRAE J. **47**(3): 52–53.

Rossier, O. and Cianciotto, N.P. (2001) Type II protein secretion is a subset of the PilD-dependent processes that facilitate intracellular infection by *Legionella pneumophila*. *Infect. Immun.* **69**(4): 2092–2098.

Rossier, O., Edelstein, P.H. and Cianciotto, N. (2002) Role of the type II protein secretion pathway in pathogenesis of *Legionella pneumophila*. In: *Legionella* (eds. R. Marre, *et al.*), Chapter 2, pp. 13–17. ASM Press, American Society for Microbiology, Washington, DC.

Rowbotham, T.J. (1980) Preliminary report on the pathogenicity of *Legionella pneumophila* for freshwater and soil amoebae. *J. Clin. Pathol.* **33**: 1179–1183.

Rusin, P.A., Rose, J.B., Haas, C.N. and Gerba, C.P. (1997) Risk assessment of opportunistic bacterial pathogens in drinking water. *Rev. Environ. Contamin. Toxicol.* **152**: 57–83.

Sabria, M. and Yu, V.L. (2002) Hospital-acquired legionellosis: solutions for a preventable infection. *Lancet. Infect. Dis.* **2**(6): 368–373.

Schlech, W.F. (1990) *Legionella* and fountains. *Lancet* **336**: 576 (letter).

Segal, G. and Shuman, H.A. (1999) *Legionella pneumophila* utilizes the same genes to multiply within *Acanthamoeba castellanii* and human macrophages. *Infect. Immun.* **67**: 2117–2124.

Shands, K. *et al.* (1985) Potable water as a source of Legionnaires' disease. *J. Am. Med. Assoc.* **253**: 1412.

Shetty, N., Srinivasan, S., Holton, J. and Ridgway, G.L. (1999) Evaluation of microbicidal activity of a new disinfectant: Sterilox 2500 against *Clostridium difficile* spores, *Helicobacter pylori*, vancomycin resistant *Enterococcus* species, *Candida albicans*, and several *Mycobacterium* species. *J. Hosp. Infect.* **41**: 101–105.

Sidari, F.P., Stout, J.E., VanBriesen, M. J. M., *et al.* (2004) Chlorine dioxide as a disinfection method for *Legionella*. *J. Am. Water Works Assoc.* **96**: 111–118.

Singapore (2001) Code of practice for the control of *Legionella* bacteria in cooling towers. Institute of Environmental Epidemiology, Ministry of the Environment, Singapore 228231. http://www.nea.gov.sg/cms/qed/cop_legionella.pdf

Squier, C.L., Stout, J.E., Krsytofiak, S., McMahon, J., Wagener, M., Dixon, B. and Yu, V.L. (2005) A proactive approach to prevention of health care-acquired Legionnaires' disease: The Allegheny County (Pittsburgh) experience. *Am. J. Infect. Contr* (In Press).

Smith, A.W. and Brown, M.R.W. (2003) Biofilms and protozoa: a ubiquitous health hazard. In: Biofilms in Medicine, Industry and Environmental Biotechnology: Characteristics, Analysis and Control (eds., Lens P., *et al.*). IWA Publishing, London.

Spain (2001) 14770 Royal Decree 909/2001 establishing the sanitary and hygiene criteria for preventing and controlling legionellosis. *The Ministry of Health and Consumer Affairs*. Palma de Mallorca, 27 July 2001.

Spitalny, K.C., Vogt, R.L., Orciari, L.A., *et al.* (1984) Pontiac fever associated with a whirlpool spa. *Am. J. Epidemiol.* **120**: 809–817.

Srikanth, S. and Berk, S.G. (1994) Adaption of amoebae to cooling tower biocides. *Microbial. Ecol.* **27**: 283–301.

Srinivasan, A., Bova, G., Ross, T., *et al.* (2003) A 17-month evaluation of a chlorine dioxide water treatment system to control *Legionella* species in a hospital water supply. *Infect. Control. Hosp. Epidemiol.* **24**(8): 575–579.

Stäuble, P., Reichlin, S., Dieterle, T., Leimenstoll, B., Schoenenberger, R. and Maritina, B. (2001) Community-acquired pneumonia – which patients are hospitalised? *Swill Med Wkly* **131**: 188–192.

Steele, T.W. and McLennan, A.M. (1996) Infection of *Terahymena pyriformis* by *Legionella longbeachae* and other *Legionella* species found in potting mixes. *Appl. Environ. Microbiol.* **62**: 1081–1083.

Steenbergen, J.N, Shuman, H.A. and Casadevall, A. (2001). *Cryptococcus neoformans* interactions with amoebae suggest an explanation for its virulence and intracellular pathogenic strategy in macrophages. *Proc. Natl. Acad. Sci.* **98**(26): 15245–15250.

Steinert, M., Birkness, K., White, E., Fields, B. and Quinn, F. (1998) *Mycobacterium avium* bacilli grow saprozoically in coculture with *Acanthamoeba polyphaga* and survive within cyst walls. *Appl. Environ. Microbiol.* **64**(6): 2256–2261.

Steinert, M., Emody, L., Amann, R. and Hacker, J. (1997) Resuscitation of viable but nonculturable *Legionella pneumophila* Philadelphia JR32 *by Acanthamoeba castellanii. Appl. Environ. Microbiol.* **63**(5): 2047–2053.

Steinert, M., Haegle, S., Skriwan, C., Grimm, D., Fajard, M., Heuner, K., Schleicher, M., Hentschel, U., Ludwig, W., Marre, R. and Hacker, J. (2002) Interaction of *Legionella pneumophila* with *Dictyostelium discoideum* (eds., Marre, R. *et al.*), Chapter 28, pp. 161–164. ASM Press, American Society for Microbiology, Washington, DC.

Stewart, P.S., Rayner, J., Roe, F. and Rees, W.M. (2001) Biofilm penetration and disinfection efficacy of alkaline hypochlorite and chlorosulfamates. *J. Appl. Microbiol.* **91**(3): 525–532.

Stone, B.J. and Kwaik, Y.A. (1998) Expression of multiple pili by *Legionella pneumophila*: identification and characterization of a type IV pilin gene and its role in adherence to mammalian and protozoan cells. Infect. Immun. **66**: 1768–1775.

Storey, M.V., Winiecka-Krusnell, J., Ashbolt, N.J. and Stenström, T.-A. (2004a) The efficacy of heat and chlorine treatment against thermotolerant *Acanthamoebae* and *Legionellae*. *Scand. J. Infect. Dis.* **36**: 656–662.

Storey, M.V., Ashbolt, N.J. and Stenstrom, T.-A. (2004b) Biofilms, thermophilic amoebae and *Legionella pneumophila* – a quantitative risk assessment for distributed water. *Water Sci Technol.* **50**(1): 77–82.

Storey, M.V., Långmark, J., Ashbolt, N.J. and Stenström, T.-A. (2004c) The fate of *Legionellae* within distribution pipe biofilms: measurement of their persistence, inactivation and detachment. *Water Sci. Technol.* **49**: 269–275.

Stout, J.E., Yu, V.L. and Best, M.G. (1985) Ecology of *Legionella pneumophila* within water distribution systems. *Appl. Environ. Microbiol.* **49**(1): 221–228.

Stout, J.E., *et al.* (1992) Potable water as the cause of sporadic cases of community-acquired Legionnaires' disease. *New Engl. J. Med.* **326**: 151.

Stout, J.E. and Yu, V.L. (1997) Legionellosis. *New Engl. J. Med.* **337**: 682–687.

Stout, J.E. and Yu, V.L. (2003) Hospital-acquired Legionnaires' disease: new developments. *Curr. Opin. Infect. Dis.* **16**(4): 337–341.

Stout, J.E. and Yu, V.L. (2003b) Experiences of the first 16 hospitals using copper-silver ionization for *Legionella* control: implications for evaluation of other disinfection modalities. *Infect. Control. Hosp. Epidemiol.* **24**(8): 563–568.

Stout, J.E. and Muder, R.R. (2004) *Legionella* in Residential Water Systems. *ASHRAE J.* **46**: 52–54.

Surman, S., Morton, G., Keevil, B. and Fitzgeorge, R. (2002) *Legionella pneumophila* proliferation is not dependent on intracellular replication (eds., Marre, R., *et al.*), Chapter 17, pp. 86–89. ASM Press, American Society for Microbiology, Washington, DC.

Sutherland, E.E. and Berk, S.G. (1996) Survival of protozoa in cooling tower biocides. *J. Ind. Microbiol.* **16**: 73–78.

Swanson, M.S. and Bachman, M.A. (2002) The *Legionella pneumophila* life cycle: Connections between growth phase, virulence expression, and replication vacuole biogenesis (eds., Marre, R., *et al.*), Chapter. 15, pp. 74–81. ASM Press, American Society for Microbiology, Washington, DC.

Swanson, M.S. and Fernandez-Moreia, E. (2002) A microbial strategy to multiply in macrophages: the pregnant pause. *Traffic* [illegible]

Swanson, M.S. and Hammer, B.K. (2000) *Legionella pneumophila* pathogenesis: A fateful journey from amoebae to macrophages. *Annu. Rev. Microbiol.* **54**: 567–613.

Texas, Department of State Health Services (1999) Report of the Texas Legionnares' Disease Task Force. www.tdh.state.tx.us/ideas/legionnaires/taskforce/report/

Thantsha, M.S. (2002) Electrochemically activated water as an environmentally safe disinfectant. M.Sc. Thesis. Department of Microbiology and Plant Pathology, University of Pretoria, Pretoria, South Africa.

The Netherlands (2000) The Temporary Mains Water (*Legionella* Prevention) Regulation. http://www2.minvrom.nl/pagina.html?id = 480

Thom, S., Warhurst, D. and Drasar, B.S. (1992) Association of *Vibrio cholerae* with fresh water amoebae. *J. Med. Microbiol.* **36**(5): 303–306.

Thomas, V., Bouchez, T., Nicolas, V., Robert, S., Loret, J.F. and Levi, Y. (2004) Amoebae in domestic water systems: resistance to disinfection treatments and implication in *Legionella* persistence. *J. Appl. Microbiol.* **97**(5): 950–963.

Tiefenbrunner, F., Arnold, A., Dierich, M.P. and Emde, K. (1993) Occurrence and distribution of *Legionella pneumophila* in water systems of central European private homes. In: Legionella*: Current Status and Emerging Perspectives* (eds. Barbaree, J.M., *et al.*) American Society for Microbiology, Washington, DC.

Tobin, J.O., Bartlett, C.L.R. and Waitkins, S.A. (1981) Legionnaires' disease: further evidence to implicate water storage systems as sources. *Br. Med. J.* 282: 573.

Walker, J.T., *et al.* (1998) Control of *Legionella pneumophila* in a hospital water system by chlorine dioxide. *J. Ind. Microbiol.* **15**: 384.

Walker, J.T., Bradshaw, D.J., Fulkford, M.R. and Marsh, P.D. (2003) Microbiological evaluation of a range of disinfect products to control mixed-species biofilm contamination in a laboratory model of a dental unit water system. *Appl. Environ. Microbiol.* **69**(6): 3327–3332.

Watnick, P. and Kolter, R. (2000) Biofilm, city of microbes. *J. Bacteriol.* **182**(10): 2675–2679.

Western Australia (2000) Code of Practice – Prevention and Control of Legionnaires' Disease. Minister for Labour Relations. www.safetyline.wa.gov.au/pagebin/pg001081.htm

Westrell, T., Schonning, C., Stenstrom, T.A. and Ashbolt, N.J. (2004) QMRA (quantitative microbial risk assessment) and HACCP (hazard analysis and critical control points) for management of pathogens in wastewater and sewage sludge treatment and reuse. *Water Sci. Technol.* **50**(2): 23–30.

WHO (World Health Organization) (1997) WHO Recommended Surveillance Standards. WHO/EMC/DIS/97.1, p. 59–60. *www.who.int/emc*

WHO (World Health Organization). (2002) *Legionella*. In: Guidelines for Drinking-Water Quality (contributors: Botzenhart, K., Bartram, J., Butler, J., Havelaar, A.H., Lee, J., Mariee, T., Pastoris, C., Plouffe, J., Tiefenbrunner, F. and Yu, V.). World Health Organization, Geneva, Switzerland.

WHO (World Health Organization). (2004) *Guidelines for drinking water quality*, 3rd Ed., Chapter 4. World Health Organization. ISBN 92 4 154638 7.

Winiecka-Krusnell, J., Wreiber, K., von Euler, A., Engstrand, L. and Linder, E. (2002) Free-living amoebae promote growth and survival of *Helicobacger pylori*. *Scand. J. Infect. Dis.* **34**: 253–256.

Winiecka-Krusnell, J. and Linder, E. (1999) Free-living amoebae protecting *Legionella* in water: the tip of an iceberg? *Scand. J. Infect. Dis.* **31**(4): 383–385.

Wireman, J.E., Schmidt, A., Scavo, C.R. and Hutchins, D.T. (1993) Biofilm formation by *Legionella pneumophila* in a model domestic hot water system. In: *Legionella*: Current Status and Emerging Perspectives. Barbaree, J.M., *et al.* (eds.). American Society for Microbiology, Washington, DC.

Woese, C.R. (1994) There must be a prokaryote somewhere: microbiology's search for itself. *Microbiol. Rev.* **58**(1): 1–9.

Woo, A.H., Goetz, A., and Yu, V.L. (1992) Transmission of *Legionella* by respiratory equipment and aerosol generating devices. *Chest* **102**(5): 1586–1590.

Yabuuchi, E. and Agata, K. (2004) An outbreak of legionellosis in a new facility of hot spring bath in Hiuga City. *Kansenshogaku Zasshi* **78**(2): 90–98.

Yee, R.B. and Wadowsky, R.M. (1982) Multiplication of *Legionella pneumophila* in unsterilized tap water. *Appl. Environ. Microbiol.* **43**: 1330–1334.

Yu, V.L. (1993) Could aspiration be the major mode of transmission for *Legionella*? *Am. J. Med.* **95**(1): 13–15.

Yu, V.L., Greenberg, R.N., Zadeikis, N., Stout, J.E., Khashab, M.M., Olson, W.H. and Tennenberg, A.M. (2004) Levofloxacin efficacy in the treatment of community-acquired legionellosis. *Chest* **125**(6): 2135–2139.

Zacheus, O.M. and Martinkainen, P.J. (1994) Occurrence of legionellae in hot water distribution systems of Finnish apartment buildings. *Can. J. Microbiol.* **40**(12): 993–999.

Zinkevich, V., Beech, I.B., Tapper, R., and Bogdarina, I. (2000) The effect of super-oxidized water on *Eschericia coli*. *J. Hosp. Infect.* **46**: 153–156.

Index

IWA Publishing's authorised EU representative for General Product Safety Regulations is Diane D'Arras, 15 rue Duret, 75116 Paris, France, e-mail: safety@iwap.co.uk.

Printed and bound by CPI Group (UK) Ltd, Croydon, CR0 4YY
12/05/2026
02108864-0019